CLIMATE,
VEGETATION & MAN

Anticyclonic Mist

In calm, anticyclonic weather early morning mist often follows a clear sky at night, when the surface of the earth rapidly cools. An *inversion of temperature* exists, and near the ground temperatures are lower than they are farther up in the air (see Chapter XIII).

CLIMATE,
VEGETATION & MAN

Leonard Hadlow, B.A.

Senior Geography Master, Burnage High School, Manchester

GREENWOOD PRESS, PUBLISHERS
NEW YORK

PREFACE

THE influence of climate upon life on the earth—plant, animal, and human—is of the highest importance and can hardly be over-emphasised. The material welfare of the various peoples of the world to a very large extent depends upon climate. Moreover, man's mental and spiritual development is, in part, controlled by the climatic. environment in which he finds himself, and in the course of history many civilisations have arisen and decayed with changing climatic conditions.

In this book an attempt has been made to survey life in its climatic setting. Part I deals with the principles that govern day and night, the seasons, and the world distribution of temperature, atmospheric pressure, winds, rainfall, and ocean currents. In Part II natural vegetation is studied in relation to its climatic needs. The major climatic regions of the world are described in Part III, special emphasis being placed upon the way in which climate governs the activities of man.

For many useful suggestions I am indebted to Miss E. M. Coulthard, Mr. R. Abbott, and Mr. W. H. Shepherd, and in particular I owe much to Mr. N. K. Horrocks for his valuable criticism and advice. I also take this opportunity of expressing my sincere gratitude to Mr. D. F. Attenborough for help generously given at all stages in the preparation of this book. Finally, to Mr. L. Brooks my special thanks are due for his encouraging guidance and for the critical care with which he edited the typescript and proofs.

L. H.

CONTENTS

Preface   7

PART I

THE FOUNDATIONS OF CLIMATE

I   The Earth in Space   11

II   Position on the Earth   20

III   The Daily Rhythm   28

IV   The Seasonal Rhythm   37

V   Climate and Weather   51

VI   Temperature and its Chief Controls   60

VII   Temperature and its Minor Controls   71

VIII   How Temperature is Mapped   78

IX   Pressure Belts   84

X   The Winds of the World   90

XI   Why it Rains   103

XII   A World Pattern of Cloud and Rain   110

XIII   War and Peace in the Air   117

XIV   The Movements of the Waters   129

CONTENTS

PART II

NATURE'S RESPONSE TO CLIMATE—NATURAL VEGETATION

XV   The Earth's Carpet of Vegetation   137

XVI   How Climate Helps to Control Vegetation   145

PART III

MAN'S RESPONSE TO CLIMATE AND VEGETATION

XVII   The Very Cold Belt   154

XVIII   The Cold Belt   165

XIX   Cool Temperate Margins   174

XX   Warm Temperate Margins   183

XXI   Cool and Warm Temperate Continental Regions—1   199

XXII   Cool and Warm Temperate Continental Regions—2   210

XXIII   The Hot Belt—1. Tropical Monsoon Lands   220

XXIV   The Hot Belt—2. Tropical Maritime Margins   230

XXV   The Hot Belt—3. Hot Deserts   236

XXVI   The Hot Belt—4. Tropical Continental Lands   250

XXVII   The Hot Belt—5. Equatorial Lands   266

XXVIII   Conclusion   279

Index   283

THE FOUNDATIONS OF CLIMATE

CHAPTER I

The Earth in Space

IMPORTANCE OF THE SUN

IN the early morning of June 29th, 1927, several thousand people poured from all parts of Britain into the little Yorkshire town of Giggleswick. All were excitedly looking forward to seeing a total eclipse of the sun, a spectacle which would not be repeated in the British Isles until 1999, when some of you may be fortunate enough to see it.

Centuries ago the inhabitants of ancient China would have beaten gongs to scare away the dragon that was devouring the sun, but we now know exactly what happens on such occasions. In its journey round the earth, the moon sometimes passes between the sun and our planet and for a few moments completely blots out all daylight.

On this June morning the long, pencil-shaped shadow of the moon swept through space and, racing across Britain, brought a temporary twilight to Giggleswick in particular. For a few minutes the attention of the whole scientific world was focused upon those places over which the shadow passed. During the eclipse huge tongues of red flame could be seen streaming out beyond the darkened rim of the sun; the temperature fell, and in the sky reappeared stars which in daylight fade away in face of the sun's dazzling competition. Scientists, who eagerly grasp such rare opportunities to unravel the secrets of the universe, therefore busied themselves with their telescopes, while excited spectators saw an unparalleled marvel of the skies. Unfortunately, in many parts of Britain the spectacle was marred by the heavy clouds so typical of a British summer.

[*Popper*

A Total Eclipse of the Sun

The faint, outer part of the film of light is called the *corona*. In the denser, inner part, or *chromosphere*, very bright tongues known as *solar prominences* are sometimes seen. Some are over 80,000 miles high.

Now, suppose that at the climax of this eclipse both earth and moon had suddenly come to a standstill, and that not merely the British Isles but the whole world had remained shrouded in darkness. Try to imagine our earth with no sunshine! It would undoubtedly be a dead world, for sunshine is essential for life and health. Plants and animals depend upon the sun for the warmth and light so necessary for growth. They also require a water supply, but without the sun there would be much less rain, if any at all. Just as the furnace beneath a boiler creates a circulation of hot water, so the sun's heat sets in motion circulations of air which result in cloud and rain. Again, when

12

we sit before a blazing fire on a cold winter day we are really deriving warmth and comfort from the sun, for coal consists of the remains of huge forests of bygone ages. These owed their existence to light and heat and, indirectly, to rain received from the sun. Coal is in truth a "carbon copy" of the sun.

THE SOLAR SYSTEM

THIS sun which maintains life and controls weather on our planet is really a star, although, being only 93,000,000 miles away, it appears so big that it does not look like one. It consists mainly of hydrogen, and contains over 300,000 times as much material as the earth. In size it is to the earth what a football would be to a pin's head.

The tiny particles of hydrogen in the sun combine to form helium and in doing so generate energy. This fusion, or union, of hydrogen particles into helium is the same process as that used in a hydrogen bomb, and the reverse of the fission, or splitting, of the atom whereby power is produced in the atomic bomb. The generation of energy produces temperatures of 6,000° C. at the surface of the sun, while at its centre the temperature is 20,000,000° C. From this solar power-house the energy streams out through space, and on reaching the earth is converted into life-giving heat. Gradually the sun is growing bigger and hotter, and eventually, more than 10,000 million years hence, it will be far too hot for life to exist on earth.

Until recently scientists believed that ages ago the tiny speck on which we ride through space was part of the sun. It was thought that a big star passed near enough to the sun to rip away from it a huge gaseous "tidal wave". From these gases slowly solidified the earth and the other planets which revolve round the sun to make up the solar system. A recent theory, however, throws doubt upon this solar origin of the earth. It suggests that probably the sun was once a "double-star", i.e. it had a partner, and that the two stars circled around each other. Eventually, with a blaze not to be equalled by the explosion of as many atomic bombs as there are dust particles in the air, the sun's "twin" burst asunder. Most of it sped away into space, but a wisp of gas remained and began to whirl

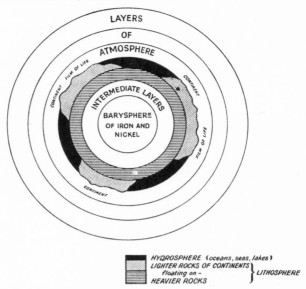

HYDROSPHERE (oceans, seas, lakes)
LIGHTER ROCKS OF CONTINENTS
        floating on –   } LITHOSPHERE
HEAVIER ROCKS

Fig. 1.—*The Concentric Layers of the Earth*
The layers are not drawn to scale; atmosphere, hydrosphere, and lithosphere are all far too wide.

around the sun. In time a few gigantic planets were condensed from some of this gas. These in their turn eventually broke up, and from the fragments were built up the earth and its fellow planets, together with their various satellites.

The earth consists of a number of concentric layers of different materials, spreading outwards from a very heavy liquid core called the *barysphere* (Greek *barus*=heavy). Most likely this core consists of heavy metal, probably iron, although one theory suggests that it is made of rocks squeezed by tremendous pressure into a metallic form. It is commonly supposed that the barysphere is extremely hot, although there are some grounds for believing that it may be no hotter than a wood fire. Surrounding the barysphere are layers of less heavy materials, which in turn give place to the *lithosphere* (Greek *lithos*=stone), or zone of rocks. The lithosphere consists of two layers. The lighter, upper one forms the continents and, like an iceberg in the sea, floats and slowly drifts, deeply sunk into a heavier and not quite solid layer below (Fig. 1). Upon the lithosphere rest the even lighter oceans and seas of the *hydrosphere* (Greek *hudor*=water). Finally, enveloping all these "skins" of the earth are the gaseous layers of the *atmosphere* (Greek *atmos*=vapour), which peters out into the thin gas, mainly hydrogen, that fills space.

THE EARTH IN THE UNIVERSE

THE solar system is but a tiny part of the universe. Some stars are so far away that to calculate their distances from the earth in miles would be as absurd as to measure Britain in millionths of an inch, and for this reason *light years* are used instead of miles. A light year is the mileage which light travels in one year —at a speed of 186,000 miles per second! Rays from some stars take millions of light years to reach us, and even from our own sun they take over eight minutes. In theory, therefore, the inhabitants (if any) of different stars, looking at our planet through telescopes of extraordinary power, would see unfolded before them the story of the human race. At the same moment some would watch Drake defeat the Armada, others would see King Harold killed at Hastings, while those even farther away would view our ancestors painting themselves with woad—and so on far back into the mists of prehistoric time!

The sun and its planets form part of the Milky Way *galaxy* —a huge, disc-shaped whirl of stars and gas. The visible part of the universe contains 100 million rotating galaxies, each made up of at least 1,000,000 planetary groups like our solar system. "There are probably as many stars as there are grains of sand on all the seashores of all the world" (Sir James Jeans).

On this vast scale our earth is an insignificant speck. Yet it is important in that conditions upon it are favourable enough to support life—although there is every reason to suppose that life exists on many other stars in the universe. Fig. 1 shows how life as we know it forms a thin layer around the earth. This film of life occurs where the lithosphere and hydrosphere meet the atmosphere, and it is thus sandwiched between the huge, heavy, lifeless core and the light, equally lifeless heights of the upper atmosphere. This zone of living matter is only about a dozen miles wide, extending from a ceiling of aerial creatures some eight miles above sea-level to a floor of marine forms little more than two miles beneath the oceans. In the following chapters we shall study this "film of life" and find out how it is largely controlled by the land, sea, and air which give it a home.

SHAPE OF THE EARTH

PRIMITIVE man, if he thought at all about the shape of the earth, probably considered it to be like a disc, for so it appears. In ancient Greece, however, wise men realised that it was spherical, but during the Dark Ages Western Europe lost touch with Greek civilisation and many people argued that the earth could not possibly be round. In fact, one monk related how he had crawled to the edge of the world and had peered over it. Even to-day there are a few people who believe that the earth is flat.

The earth was regarded as the centre of the universe. Around it spun concentric transparent spheres in which were embedded the stars. Each whirling sphere gave out a musical note and, since Nature could not tolerate discord, the various notes combined to create the harmonious "music of the spheres". Writing in the seventeenth century, Sir Thomas Browne says, "There is a music of the spheres, though they give no sound unto the ear, yet to the understanding they strike a note most of harmony."

So many absurd travellers' tales were told in those days that it is not surprising that sailors dared not, for fear of the unknown, sail too far out of sight of land. Eventually the discoveries of Copernicus, Galileo, Kepler, and Newton paved the way for modern ideas concerning the earth and its place in the universe.

It is now known that the earth is shaped almost like a ball. It is slightly flattened at the Poles, and here and there it probably bulges. These irregularities are so small that for practical purposes they may be ignored. The presence of mountains makes impossible a truly spherical surface, but such "bumps", impressive though they appear to us in their grandeur, are really negligible. On a globe of about one foot diameter the highest mountains could be represented by the thickness of a coat of varnish, while if water is poured over it and allowed almost to

U.S. Information Service

The Curvature of the Earth

Photographed from a rocket at a height of 100 miles over the United States and Mexico, the curvature of the earth can clearly be seen. The dark inlet is the Gulf of California. The area shown is nearly twice that of the British Isles.

evaporate the moist film would give a correct impression of oceanic deeps. How insignificant, on this scale, must man appear as he crawls about the earth! It has been said that the human race could be packed into a box with sides half a mile in length, which, if tossed into the Grand Canyon of the Colorado River, would be invisible on the earth as a whole.

17

How can we prove that the world is round? When climbing a mountain, we notice that the horizon slowly widens. Landscapes still hidden by the curvature of the earth from people farther down the mountain-side gradually come into view, whereas over a flat earth the horizon at the summit would be no more extensive than that at the base. From an aeroplane infra-red ray photographs have been taken, showing the coast-line of South-east England from the Isle of Wight to the Wash. The curvature of the surface has, in fact, been revealed by an automatic camera carried aloft by a rocket to a height of over ninety miles. Again, a lunar eclipse occurs whenever the moon passes into the earth's shadow, which is then seen to be round. A sphere is the only body that, no matter what position it takes up, always casts a round shadow.

SIZE OF THE EARTH

THE greatest possible circumference around the earth is approximately 25,000 miles. Owing to the flattening of the surface at the Poles, the diameter varies. The polar diameter, or axis, is

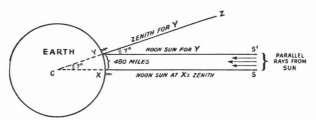

Fig. 2.—How the Earth's Circumference is Measured

7,900 miles long and the equatorial diameter 7,926. To calculate these distances, it is not necessary to circumnavigate the globe trailing a tape-measure.

In Fig. 2 an observer, X, is shown and another, Y, stands several hundred miles due north of him. For X the noon-day sun is overhead (the ray SX, if produced, would therefore continue to C, the centre of the earth). To find his zenith, or overhead position, Y looks in the direction of the line CYZ, but does not see the sun there. To do so he must look along the line YS', parallel to XS, for all rays from the sun received by the

18

earth are, for all practical purposes, parallel and should always be drawn so in diagrams (Fig. 9). The angular distance by which the sun departs from Y's zenith ($\angle Z Y S'$) can be measured by a sextant, an instrument used in navigation to find the position of a ship or aeroplane.

Since CYZ is a straight line crossing the two parallel lines YS' and CXS, $\angle Z Y S' = \angle Y C X$.

Suppose that $\angle Z Y S'$ (and therefore $\angle Y C X$) is found to be 7°. The mileage from X to Y (i.e. the arc XY) is now very carefully measured. Suppose that it is 480 miles.

An arc of 480 miles corresponds to 7° at the centre.

$\therefore$ An arc of $\dfrac{480}{7}$ miles corresponds to 1° at the centre.

$\therefore$ The whole circumference corresponding to 360° at the centre of the earth $= \dfrac{480}{7} \times 360 = 24{,}686$, or nearly 25,000 miles. To be exact, the equatorial circumference measures 24,902 miles.

The first man to carry out this experiment was Eratosthenes (born 276 B.C.), the Greek custodian of a famous library at Alexandria in Egypt. His result was not quite accurate, but his efforts deserve our admiration, considering how crude, judged by modern standards, were his instruments.

EXERCISES

1. Find out the difference between planets, asteroids, comets, meteors, meteorites, stars, and galaxies.

2. Alpha Centauri, the nearest star to the earth, is 25,000,000 million miles away. If it suddenly ceased to exist, for how long (to the nearest year) would it continue to shine, assuming that light travels at about 16,000 million miles per day?

3. How long would it take an aeroplane travelling at the speed of sound (750 miles per hour) to reach (a) the moon (240,000 miles away) and (b) the sun (93,000,000 miles away)?

CHAPTER II

Position on the Earth

MAKE a chalk dot on a football, and then try to describe its exact position. It is no easy task. In similar problems on the earth, the position of places can be determined by making use of certain lines and angles.

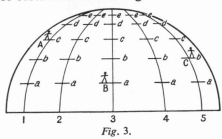

Fig. 3.

Suppose that three men, A, B, and C, are cleaning an enormous glass dome and are standing upon curving ladders which, clamped tightly to it, meet at the top (Fig. 3). The ladders are numbered 1, 2, 3, 4, etc., and their rungs are lettered, a, b, c, d, e, etc. The position of A can now be described as being on ladder 1, rung c (or 1c). Similarly, B is at 3a and C at 5b.

A globe is covered by a somewhat similar pattern of curving lines. Corresponding to the uprights of the ladders are semi-circles called meridians of longitude, while the rungs are repre-sented by circles known as parallels of latitude. If both the meridian and the parallel on which a place is situated are known, then the point at which they cross each other fixes its position.

In this chapter we shall consider six different kinds of lines and two angles. The lines are drawn upon maps, but over the surface of the earth they are quite imaginary. Only in a few special cases is their presence sometimes indicated on the ground. For instance, near Quito in Ecuador the Equator is marked by a monument, while at a boarding school for native boys in Kenya it is taken as the half-way line on the football pitch. In Finland a signpost points to the Arctic Circle, and at the Greenwich Royal Observatory the Prime Meridian is for

20

The Equator

[*Popper*

This signpost in Kenya marks the position of the Equator.

a few yards shown by a stone strip. Golfers in São Paulo (Brazil) are told that the Tropic of Capricorn passes through the municipal golf-course, so that they can drive a golf-ball from the tropics into the temperate zone and vice versa. Again, on "crossing the line" (the Equator) for the first time passengers on board ship sometimes take part in the ceremony of paying tribute to King Neptune.

The Six Lines

1. *Axis of Rotation.*—Round this line, which is 7,900 miles long, the earth rotates once in every twenty-four hours. Its ends are called the North Pole and South Pole. Explorers know when they reach the Pole only by finding the angular height of the noonday sun above the horizon and then making certain calculations. It is a mistake to think that the compass needle points towards this geographical North Pole; in the northern hemisphere, for instance, it turns towards a slowly moving magnetic North Pole, situated at present in North-eastern Canada.

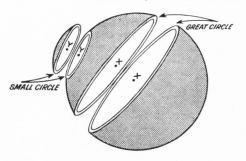

Fig. 4.—*Great and Small Circles*

X = centre of Great Circle (also earth's centre).
Y = centre of Small Circle (not earth's centre).

2. *Great Circle.*—Cut an orange into halves. Note that the rim of each half forms the greatest possible circumference and that its centre is also the centre of the orange. In exactly the same way, a great circle is any circle so placed round the earth that a huge knife, slicing along it, would cut the world into halves, or hemispheres. The centre of a great circle, therefore, coincides with that of the earth (Fig. 4). There is no limit to the number of such circles that can be drawn, and they are all approximately 25,000 miles long.

Over a globe tightly pull a piece of string between London and Vancouver (British Columbia), so that it shows the shortest route between them. If the string is now stretched in the same direction round the whole globe, you will notice that it forms a great circle. The shortest route between two places is therefore provided by the great circle that passes through them, and if it can be followed by ships and aircraft much time and fuel will be saved. Unfortunately, navigation along great circle routes is not always practicable.

The shortest route between two widely separated places sometimes takes a surprising course. For instance, the route you have marked between Britain and Western Canada leaves London via Glasgow in a N.W. direction, skirts the south of Iceland and passes westwards over Greenland before turning S.W. over Northern Canada and approaching Vancouver going in a S.S.W. direction (Fig. 5). A great circle therefore constantly changes direction.

On a Mercator map of the world draw a straight line between the same two cities (Fig. 5). Apparently this line indicates the shortest distance between them, but contrast it with the great circle route on the globe and you will notice that the two routes

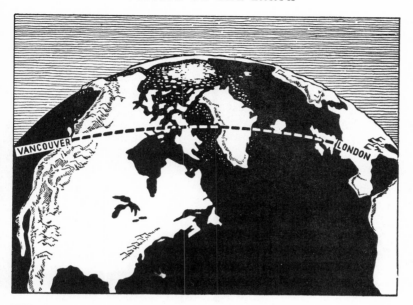

KEY: ✦•▪▪◂ Great Circle (i.e. shortest) route from London to Vancouver.
——— Shortest route according to Mercator map.

Fig. 5.—*The Great Circle Route from London to Vancouver as it appears on the Globe (above) and on a Mercator Map (below)*

are quite different. You will now realise that maps are mis-leading; in fact, all maps are incorrect in one way or another, for the curved surface of the earth cannot without distortion be represented on a flat chart.

3. *Small Circle.*—A small circle is one whose centre does not coincide with that of the earth and whose plane divides the world into two unequal portions (Fig. 4). A small circle route between two places cannot be the shortest one.

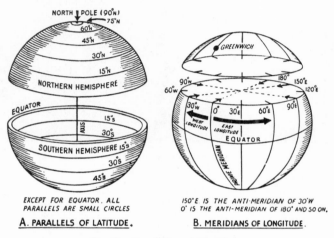

Fig. 6.—Parallels and Meridians

4. *Equator.*—The Equator is the great circle midway between the Poles; its plane is at right angles to the axis and divides the world into northern and southern hemispheres (Fig. 6A).

5. *Parallel of Latitude.*—Parallels (the "rungs" of the "ladders" by which positions on a map are fixed) are small circles whose planes are at right angles to the axis. They are therefore parallel to the Equator, which is unique in that it is the only parallel which is a great circle (Fig. 6A). For this reason the Equator provides an obvious starting line from which all other parallels are numbered. From 0° (the Equator) they range northwards and southwards to a maximum of 90° N. at the North Pole

and 90° S. at the South Pole. Certain parallels have names, e.g. the Equator, the Tropics of Cancer (23½° N.) and Capricorn (23½° S.), and the Arctic and Antarctic Circles (66½° N. and 66½° S.). In North America the 49th parallel forms a large part of the boundary between Canada and the United States.

What do you notice about the length of parallels between the Equator and the Poles?

6. *Meridian of Longitude.*— Crossing the "rungs" of our imaginary "ladders" are their curving uprights, known as meridians. A meridian is a semi-great circle which links the North and South Poles.

[*Popper*

The Prime Meridian

This stone slip at Greenwich marks the Prime Meridian.

Directly opposite to any given meridian is its anti-meridian, and together the two make up a complete great circle (Fig. 6B). Unlike parallels, all meridians are equal in length, so that no single one can be chosen as a natural starting line, or *prime meridian*, from which to number the rest. A prime meridian (0°) must therefore be artificially selected, and from it the others are numbered eastwards and westwards to a maximum of 180° each way. All meridians, save the prime meridian and the 180th, are labelled either east or west. The meridian of 180° forms the anti-meridian to the prime meridian, 90° E. is opposite to 90° W., 30° W. to 150° E., and so on. Notice that the sum of a meridian and its anti-meridian is always 180°. Since 1883 most countries have recognised as the prime meridian the one that passes through Greenwich, near London.

Two Important Angles

1. *Latitude.*—In Fig. 7 two imaginary tunnels are shown burrowing through to the centre of the earth, one from London and the other from the point X where London's meridian crosses the Equator. The angle that they make at the centre of the earth is called the latitude of London north of the Equator. It amounts to $51\frac{1}{2}°$ N. Latitude is therefore the angular distance, measured at the centre of the earth, of a place north or south of the Equator. In each hemisphere the Pole marks the maximum latitude of 90°.

The arc of the earth's surface between the mouths of the

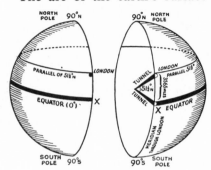

Fig. 7.—Latitude

tunnels at London and at X, corresponding to $51\frac{1}{2}°$ of latitude, would measure 3,565 miles. Therefore the surface distance corresponding to 1° of latitude is $3,565 \div 51\frac{1}{2} = 69\cdot2$, or approximately 69 miles.

For accurate measurement of latitude, each degree (°) is divided into sixty equal minutes (')

and each minute into sixty equal seconds ("). The surface length corresponding to 1' of latitude is therefore $\frac{69}{60}$ miles. Sailors call this distance a nautical mile, and a *knot* is a speed of one nautical mile per hour.

If, pivoted upon the centre, the tunnels are swivelled round the earth in an east-west direction or vice versa, their mouths will trace out upon the surface two parallels, one being the Equator and the other the parallel of $51\frac{1}{2}°$ N.

A parallel therefore connects all places of the same latitude and should not be confused with latitude itself.

2. *Longitude.*—New York is approximately on parallel 40° N., but its exact whereabouts on this circle can be found only if we know its longitude. In Fig. 8 are shown two imaginary tunnels boring in along the plane of parallel 40° N. to the axis

and not, as in latitude, to the centre of the earth. One runs from New York itself and the other from X, the point where the prime meridian crosses parallel 40° N. The angle that they make at the axis is called the longitude of New York west of the prime meridian. It amounts to 74° W. Longitude is thus the angular distance, measured at the axis, of a place east or west of the prime meridian.

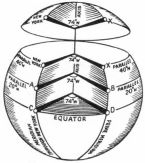

Fig. 8.—Longitude

The maximum longitude is 180° eastwards and westwards, i.e. a complete circle of 360° in all. Like the prime meridian, the 180° meridian is labelled neither east nor west, but is a boundary between the eastern and western hemispheres.

The distance between the mouths of the tunnels at New York and at X, corresponding to 74° of longitude, would measure about 3,900 miles. The arcs between similar tunnels (i) at A and B, and (ii) at C and D, also corresponding to 74° of longitude but on parallel 20° N. and on the Equator respectively, would be about 4,800 and 5,100 miles. Degrees of longitude, therefore, unlike those of latitude, are bounded at the earth's surface not by a constant distance but by arcs decreasing from approximately 69 miles along the Equator to nothing at the Poles.

A meridian connects all places of the same longitude, and should not be confused with longitude itself.

EXERCISES

1. With chalked string, mark on a globe the great circle route between pairs of widely separated cities. Contrast each one with what seems in an atlas to be the shortest route.

2. Suggest reasons why ships and aeroplanes do not always follow great circle routes.

3. The antipodes of a place is the point which is diametrically opposite to it on the world. Find in an atlas which place most nearly marks the antipodes of: Gibraltar (36° 10′ N., 5° 20′ W.), Perth (32° 0′ S., 115° 50′ E.), Tientsin (39° 0′ N., 117° 5 E.) and Cherbourg (49° 39′ N., 1° 40′ W.).

4. Find the speed in knots (a) of a ship which sails through 5° of latitude in one day, keeping to the same meridian, and (b) of one which sails through 8° of longitude in one day along parallel 60° N., which is half the length of the Equator.

CHAPTER III

The Daily Rhythm

FAR from remaining fixed in the universe, the earth is rapidly moving. Several different movements are made, but in the study of geography two only are of outstanding importance. From each one springs a well-marked rhythm which controls our activities, our clothing, the food we eat, and many other aspects of our daily life.

The rotation of the earth around its axis, the movement which forms the subject of this chapter, leads to three very important results:

1. Day and Night.
2. Time and its measurement.
3. The deflection of moving bodies, summed up in Ferrel's Law.

ROTATION AND THE RHYTHM OF DAY AND NIGHT

IMAGINE what life would be like without the alternation of day and night. During the hours of daylight man, conscious and alert, is busily occupied with the task of earning his living, while night is equally essential as a time for rest and sleep. What causes this all-important rhythm of day-night-day-night?

Suppose that we are in a train approaching a long tunnel. Darkness seems to rush towards us, and once we have entered the tunnel daylight rapidly retreats behind us. For a while all is pitch-dark but at length, far ahead and apparently advancing quickly towards us, appears a circle of daylight, and as we emerge from the tunnel night speeds away to the rear. If the track formed a circle, as in a toy railway, we should experience a definite rhythm of light-dark-light-dark as we went in and out of the tunnel.

So it is with daylight and darkness over the earth, with dawn,

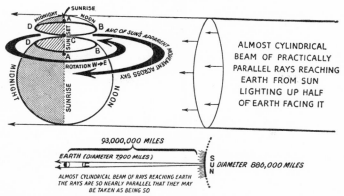

Fig. 9.—*Day and Night (above).   Why Sunlight can be said to reach the Earth as Parallel Rays (below)*

noon, dusk, and midnight. Since the earth is a sphere and the sun's rays received by it are parallel, exactly half of it will face the sun and will be lit up. The other half will be in a dark shadow, which can be likened to the railway tunnel (Fig. 9).

If the earth were perfectly still, all places would remain either in the light or in the dark half according to whether or not they faced the sun. The earth, however, is spinning like a top, and every parallel twists round from west to east, taking all the places which it connects for a circular ride.

In our example of the circular railway, it is as though the track itself raced through the tunnel into daylight. Similarly, each place on earth is carried by its rotating parallel from dawn through daylight, into darkness at sunset, through the "tunnel" of night, and out again into light at sunrise (Fig. 9).

Again, just as in our moving train we seem to stay still while the tunnel appears to travel towards us, so we on the earth, although in reality moving from west to east, seem to remain stationary while both day and night appear to come upon us from the opposite direction. The sun "rises" from an easterly direction as we spin round towards it from the west (Fig. 9, position A). It wheels across the sky in a great arc, reaching its highest point at noon as our meridian moves round to face it squarely (position B). It sinks to "set" in a westerly direction as we retreat eastwards from it to re-enter the "tunnel" (position C). Half-way through the "tunnel", when our meridian turns its back on the sun, midnight occurs (position D).

From Fig. 9 it would seem that everywhere daylight equals darkness in length, but such is the case only around March 21st and September 23rd, for reasons which will be discussed in the next chapter.

ROTATION AND TIME

BUT for the earth's rotation clocks would be useless. In spinning round, the earth acts as a master-clock by which all man-made timepieces are regulated.

On a roundabout you pass a given point on the ground at regular intervals. Similarly, if on the rotating earth a given meridian is in a certain position in relation to the sun (e.g. opposite to it at noon), it will at regular intervals come round to that same position. This interval is called a day, and for convenience it is divided into hours, minutes, and seconds. Millions of years hence the day will be considerably longer than it now is, for the drag of tides in shallow seas is gradually slowing down the earth's rotation. The moon, which is largely responsible for the tides, is thus holding back the earth with a watery "rope".

Local Sun-time.—Fig. 9 shows that sunrise (at A), noon (B), sunset (C), and midnight (D) cannot occur everywhere at the same time. At any given moment, the time of day over the world will vary according to the particular positions in relation to the sun taken up by meridians, i.e. sun-time varies with longitude. Since each place rotates through a circle of 360° of longitude in a day of 24 hours, it will move through 15° in one hour, or through 1° every four minutes.

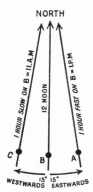

Fig. 10.— *Time changes with Longitude*

If A (Fig. 10) lies 15° east of B, where it is noon, then it has gone farther through daylight than B. Noon at A occurred one hour ago, and its local time by the sun is therefore 1 p.m. Similarly, if C is 15° west of B it has advanced through daylight less than B, and its local sun-time is one hour short of noon, i.e. 11 a.m. Time differences between places may be easily remembered as follows: If A is eAST of B, then A's clocks are fAST on B's, whereas if A is *W*est of B, then A's clocks are slo*W* on B's.

The following example illustrates the effect of this change in time of one hour with every 15° increase in longitude:

The score in a Test match played in Australia is broadcast from Sydney (150° E.) at 6 p.m. on Monday, December 4th. When will the score be heard in London (0°)?

Longitude difference between Sydney and London = 150°.

∴ Time difference between Sydney and London = $\dfrac{150}{15}$

= 10 hours.

London, being west of Sydney, lags behind it in local time, and clock time there is ten hours slow on Sydney's 6 p.m., i.e. it is 8 a.m. The score, broadcast from Australia on a summer evening, is heard in Britain over a winter breakfast-table on the same day—apparently before play has even begun! Western Canada receives the result late on Sunday night, December 3rd, i.e. the day before the match is played!

Describing a voyage across the Atlantic, Mark Twain says: "We had the phenomenon of a full moon located just in the same spot in the heavens at the same hour every night. The reason of this singular conduct on the part of the moon did not occur to us at first, but it did afterward when we reflected that we were gaining about twenty minutes every day, because we were going east so fast—we gained just about enough every day to keep along with the moon. It was becoming an old moon to the friends we had left behind us, but to us Joshuas it stood still in the same place, and remained always the same."

Standard Time.—Since each meridian represents a different sun-time, Manchester ($2\frac{1}{4}$° W.) lags nine minutes (i.e. $2\frac{1}{4}\times4$) behind Greenwich in local time. When Greenwich clocks register noon the time at Manchester should be 11.51 a.m., at Edinburgh ($3\frac{1}{4}$° W.) 11.47 a.m., at Belfast (6° W.) 11.36 a.m., and at Yarmouth ($1\frac{3}{4}$° E.) 12.7 p.m.

Imagine how difficult it would be to make a railway or airservices time-table, or a wireless programme, if the correct local time had to be considered when calculating exactly when a train, aeroplane, or broadcast feature would reach different towns. To reduce such chaos to order, the whole of a given area, known as a *Time Zone*, keeps the time of one selected meridian. This particular time thus becomes the *Standard Time* for the zone.

Time Zones.—It would be absurd if Greenwich standard time had to serve the whole world, since twelve o'clock noon by local clocks would then herald a rising sun in New Orleans (90° W.) but would bid farewell to a setting one in Calcutta (90° E.).

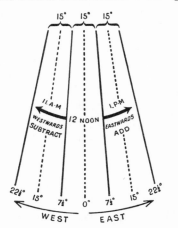

Since 1883 a system of time zones has been gradually accepted throughout the world. The 360° of longitude have been divided into twenty-four equal zones. Each one is therefore 15° wide, differs from its neighbours by one hour, and keeps as its standard time, that of its central meridian (Fig. 11).

Time-zone boundaries occasionally deviate from the selected meridians,

········· CENTRAL MERIDIAN, FROM WHICH STANDARD TIME FOR EACH ZONE IS SET.

——— MERIDIAN BOUNDARIES OF TIME ZONES

Fig. 11.—Time Zones and Standard Time

zigzagging to meet local needs, e.g. to give islands the same time as their neighbouring mainland. In Britain, a small area entirely within one time zone, no difficulties arise, but some huge countries cover more than one zone. For instance, Canada is divided into six zones, and watches must be altered when travellers pass from one into the next.

The International Date Line.—When local time at Greenwich is 12 noon Monday, at 179° W. it is just after midnight Sunday (actually 12.4 a.m. Monday), while at 179° E. it is nearly Tues-

32

day (actually 11.56 p.m. Monday). Taking the smaller longitude difference between them, 179° W. is only 2° away from 179° E. (Fig. 12), yet in time-keeping they differ by almost a day. Midway between the two, at 180°, it is midnight. Arriving at 180° westwards from Greenwich, a man would declare that it was midnight Sunday (i.e. twelve hours slow on Greenwich time), but another man arriving there eastwards from Greenwich would disagree and would insist that the time was midnight Monday (i.e. twelve hours fast on Greenwich time).

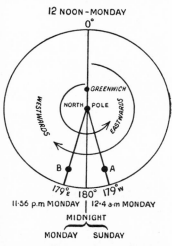

Fig. 12.—*The International Date Line*

When travellers cross this 180° meridian the date must therefore be changed. Fig. 12 shows that as this *International Date Line* is crossed westwards, e.g. from A to B, midnight Sunday advances to midnight Monday and the whole of Monday is omitted— "going west, a day 'goes west'". Going eastwards, e.g. from B to A, midnight Monday reverts to midnight Sunday and the whole of Monday is repeated.

This practice, however, is not carried out on Christmas Day, otherwise travellers who crossed the line eastwards would enjoy a double Christmas at the expense of those who, westwards bound, would miss it altogether. The same custom applies to Sundays.

The International Date Line follows the meridian of 180° across the Pacific Ocean for most of its length. Where, however, Siberia and groups of islands, e.g. Fiji, straddle the meridian, the Date Line bends away from it and keeps to the ocean, leaving the land on whichever side is politically or economically most convenient.

ROTATION AND FERREL'S LAW

A POINT on the rim of a revolving bicycle wheel travels through a larger circle than one on the hub, yet both points complete a

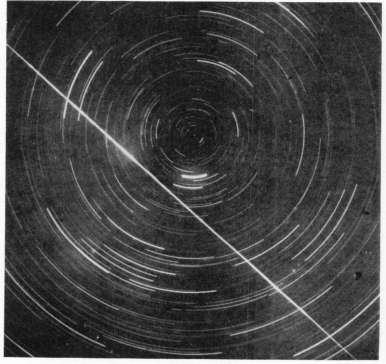

[*Popper*

The Apparent Movement of the Stars

The earth's rotation causes the stars *to appear* to move in circles. To take this photograph a camera was pointed towards the Pole Star and an exposure made of two hours. Each arc of light represents one-twelfth of the daily revolution of a separate star. The straight bright line shows the path of a meteor which crossed the sky.

revolution in exactly the same time. The rim, therefore, moves round faster than the hub; hence towards the rim the spokes appear blurred, whereas near the hub each spoke can be clearly distinguished.

Similarly, the earth's rotation gives to any place upon its surface a velocity which depends upon the latitude of the place in question. Velocity means the rate of movement *in a certain direction*—in this case from west to east. While a point on the Equator moves through 25,000 miles in 24 hours, or at over

1,000 miles an hour, a point on parallel 60° N., which is half the length of the Equator, moves through 12,500 miles at about 500 miles an hour. In the same time the North Pole merely twists round on the same spot, i.e. its velocity is nil. Rotational velocities therefore decrease from Equator to Poles.

It is difficult to realise that the British Isles, for instance, are hurtling round from west to east at velocities along different parallels ranging from over 500 miles an hour in Northern Scotland to nearly 700 miles an hour in Southern England. We fail to notice this movement because everything about us, including the atmosphere, is whirling round with us.

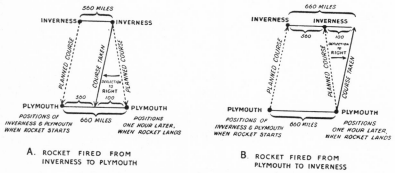

Fig. 13.—Ferrel's Law (Northern Hemisphere)

By an optical illusion the earth's west to east rotation causes the sun and stars to appear to wheel across the sky in huge arcs from east to west. Their apparent westward motion is just as fast as our eastward rotation. They do not seem to move at our rapid velocities because they are so very far away, just as an aeroplane high in the sky appears to crawl compared with its speed when, a few feet above the ground, it skims over our heads.

The changes in velocity according to latitude have an important effect on the course taken over the earth's surface by a freely moving body.

Suppose that in one hour a rocket unhampered by winds could be fired from Inverness to Plymouth, i.e. from where the eastward velocity is 560 to where it is 660 miles an hour.

Throughout its flight the projectile maintains, in addition to its southward speed, *approximately* the same easterly velocity as its point of departure, i.e. 560 miles an hour (Fig. 13A). The ground beneath it, however, is slipping away eastwards at ever-increasing speeds, and by the end of the hour's journey Plymouth has moved 100 miles (i.e. 660—560) farther east than the rocket, which consequently lands to the west of the city. There has therefore been a deflection to the *right* of the north-south course originally planned.

If a rocket is now fired in the opposite direction, from Plymouth to Inverness, it will land 100 miles to the east of its destination, since throughout its flight it maintains, in addition to its northward speed, *approximately* the same eastward velocity as its point of departure (660 miles an hour), whereas the ground beneath it is lagging behind. Again deflection has been to the right (Fig. 13B).

Between two places in the southern hemisphere it can be similarly proved that the rocket would be deflected to the *left* of the course proposed.

All moving bodies, including winds and ocean currents, are affected in this way by the earth's rotation. *Ferrel's Law* summarises the matter by stating that all freely moving bodies are deflected to the right in the northern and to the left in the southern hemisphere. You can easily remember that noRth contains R (for Right), but not L (for Left.)

EXERCISES

1. An aeroplane takes off at dawn to fly westwards round the world at 250 miles an hour along parallel 80° N., where the earth's rotation speed eastward is only 180 miles an hour. Describe what the pilot will notice about the sun's path across the sky, noting particularly in which direction he will see sunset and sunrise.

2. Using your atlas, find a city where local sun-time is approximately (*a*) 9.20 p.m. (in Japan), (*b*) 7.16 a.m. (Canada), (*c*) 8 a.m. (Brazil), (*d*) 12 noon (Gold Coast), and (*e*) 2 p.m. (Russia and Egypt) when Greenwich time is 12 noon.

3. What is the time and day (*a*) in Mexico City (100° W.) when it is 6.15 a.m. New Year's Day in Madras (80° E.), and (*b*) in Wellington (175° E.) when it is 11.30 p.m. on Christmas Eve at New Orleans (90° W.)?

4. Under what circumstances could you (*a*) celebrate a double birthday in one year, and (*b*) have no birthday at all?

CHAPTER IV

The Seasonal Rhythm

BESIDES rotating round its axis, the earth revolves round the sun, tracing out as it does so an almost circular path, or *orbit*. The sun's position within this orbit is not quite central, so that in December the earth is about 3,000,000 miles nearer to the sun than in June. Our average distance from the sun is 93,000,000 miles. The earth's speed as it races round its orbit surpasses the powers of imagination. Travelling at nearly 70,000 miles an hour, we have been hurled through space as far as Land's End is from John o' Groats before we can count one hundred—as well as being whirled round the earth's axis at speeds which in Britain range from 500 to over 600 miles an hour!

Throughout this revolution the plane of the Equator is tilted across the plane of the orbit at an angle of $23\frac{1}{2}°$, so that at any point in the journey the position taken up by the Equator is parallel to that assumed at all other points. In other words, the earth does not wobble round the sun, but revolves with its axis constantly pointing in a fixed direction at an angle of $66\frac{1}{2}°$ (i.e. $90°-23\frac{1}{2}°$) to the plane of the orbit (Fig. 14).

From this movement arise four important results:
1. A measurement of time.
2. The seasons.
3. The annual "swing" of the sun.
4. A seasonal variation in the length of daylight.

REVOLUTION AND TIME

WE have seen how rotation as a time-keeper provides us with the day. Revolution similarly fixes the year, which is the time taken by the earth to complete its roughly circular voyage round the sun. Since $365\frac{1}{4}$ rotations are made during this journey, the year consists of $365\frac{1}{4}$ days. For convenience, a calendar year

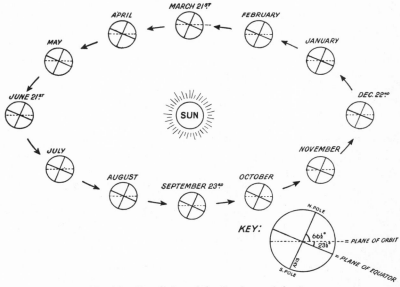

Fig. 14.—*Revolution of the Earth round the Sun*

is reckoned as 365 days, the quarter-day being ignored for three years out of every four. To put matters right, a whole day is added to every fourth year, which is called leap year.

REVOLUTION AND THE SEASONS

ALL places except those near the Equator undergo a rhythmical change of temperature as season follows season year after year. After winter, "when icicles hang by the wall," life revives in spring:

> *When daisies pied, and violets blue,*
> *And lady-smocks all silver-white,*
> *And cuckoo-buds of yellow hue,*
> *Do paint the meadows with delight.*

Spring in its turn makes way for "all the delights of summer weather". Then comes autumn, "season of mists and mellow fruitfulness", which fades into winter:

> *When the soundless earth is muffled,*
> *And the cakèd snow is shuffled*
> *From the ploughboy's heavy shoon.*

38

Finally, with the arrival of another spring the cycle is completed. For the farmer the seasons determine seed-time and harvest; the schoolboy enjoys his longest holiday in summer, when he derives greater benefit from fresh air and sunshine than is possible in the fogs and overcast skies of winter. In sport, summer means the cricket field or tennis court, and winter the more vigorous exertions of football or hockey. What we wear and the kind of food we eat are also partly determined by the seasons.

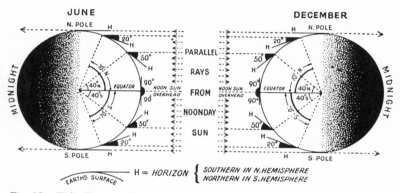

Fig. 15.—This diagram shows that there would be no differing seasons if the equatorial plane remained level with the plane of the orbit. The conditions shown do actually occur on March 21st and September 23rd, when the noonday sun is overhead at the Equator.

SUMMER AND WINTER

WHY is summer heat followed by winter cold? Fig. 15 shows the June and December positions of the earth in its journey round the sun as they would appear if the equatorial plane remained constantly level with the plane of the orbit instead of being tilted across it. Notice that:

(*a*) To an observer on the Equator the noontide sun in both months (and, in fact, throughout the year) would appear overhead.

(*b*) At every other parallel the noonday sun-angle, i.e. the angular height of the sun at noon above the observer's horizon, would not vary throughout the year. For instance, it would be 50° at parallels 40° N. and S. and 20° at 70° N. and S., while at both Poles the noonday sun would be level with the horizon.

A sun which thus climbed each day to exactly the same noonday summit would lead to no seasonal changes of temperature. Britain, for instance, would enjoy a perpetual spring, with a noon sun-angle of between 30° and 40°. The tilt of the earth, however, upsets this simple pattern of a single season.

In a darkened room hold a book open in the rays of a torch in such a way that the light rays reach the page at a low angle (Fig. 16).

BOOK AT LOW ANGLE - LIGHT ON PAGE DIM   BOOK AT HIGH ANGLE - LIGHT CONCENTRATED

Fig. 16.

Spread over a large area, the light is so dim that you cannot read the print. Now tilt up the book so that the rays reach it more directly. Exactly the same amount of light leaves the torch, but it now meets the paper at a higher angle and is concentrated upon a smaller area, making it possible to read the book. The more the surface is tilted up towards the light, the brighter does it grow.

If the light from the torch and the page of the book represent respectively the sun's rays and the earth's surface, this experiment illustrates how solar rays, by striking the earth at different angles through the year, give rise to the changing seasons. Look again at Fig. 14 and notice the following points:

(*a*) For six months, from March to September, the northern hemisphere is tilted towards the sun, while the southern hemisphere is tilted away from it. Over any given area north of the Equator the sun's noontide rays, reaching the surface at a high angle, are concentrated. For the northern hemisphere it is the summer half-year. South of the Equator beams of solar rays at noon strike the surface at lower angles than in the northern hemisphere; their heat is dispersed over greater areas and winter reigns.

40

(*b*) For the next six months, from September to March, these conditions are reversed. Rays reach the northern hemisphere, now turned away from the sun, at low angles compared with those prevailing south of the Equator, where the earth's surface is tilted well up into sunlight. Northern winter corresponds to southern summer.

(*c*) On two dates only, June 21st and December 22nd, is the whole of the axis directly in line with the sun. On June 21st the northern hemisphere therefore views the sun from a completely "full face" position, and noonday sun-angles reach their maximum values. The southern hemisphere is turned "full face" away from the sun, which here reaches its lowest noon angles. June 21st will be midsummer day north of the Equator but midwinter day south of it.

(*d*) December 22nd sees the tables turned. Southern latitudes now take up a full-face position towards the sun, while the northern hemisphere is turned directly away from it. December 22nd therefore marks southern midsummer but northern midwinter.

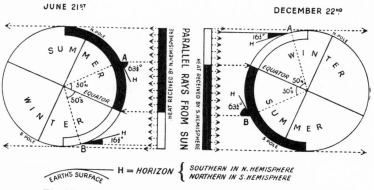

Fig. 17.—*The Tilting of the Earth's Axis causes the Seasons*

Fig. 17 shows the June 21st and December 22nd positions of the earth in its orbit. Two places, A and B, illustrate in each hemisphere the seasonal changes in the noonday sun-angle. A is at 50° N., i.e. the latitude of Land's End, while B is at 50° S., approximately the latitude of the Falkland Islands.

June 21st.—The sun-angle at A is high (63½° above the

southern horizon), whereas at B it is low ($16\frac{1}{2}°$ above the northern horizon). Heat is concentrated at A but dispersed at B. The northern hemisphere also receives a larger share (the shaded part) than the southern hemisphere of the total amount of heat that reaches the earth.

December 22nd.—The sun-angle at A is now low ($16\frac{1}{2}°$), whereas B now basks in a high noonday sun, $63\frac{1}{2}°$ above the horizon. Heat is therefore dispersed at A but concentrated at B. The northern hemisphere now receives a smaller share (unshaded part) of the sun's heat than the southern hemisphere.

Notice that on both dates, and throughout the year, the noonday sun is always high in equatorial latitudes, where the annual range of temperature is very small.

Northern seasons are therefore the reverse of those south of the Equator. As the Englishman looks forward to an (occasionally) old-fashioned snowy Christmas, the South African eats his Christmas dinner in a heat wave. The Australians visit England to play Test matches from May to August, while the English team makes a tour "down under" from November to March.

Spring and Autumn

Since the high noontide sun-angle at A on June 21st ($63\frac{1}{2}°$) gradually sinks to a low one on December 22nd ($16\frac{1}{2}°$) and vice versa at B, there must be a midway stage between these dates, i.e. September 23rd, when a balance between these angles is reached. On this day at both A and B the sun-angle is 40°, i.e. $(63\frac{1}{2}° + 16\frac{1}{2}°) \div 2$. A half-way stage between summer and winter, namely autumn, has been attained at A, and between winter and summer, namely spring, at B.

The six months from June 21st to December 22nd, however, take the earth only half-way round the sun. During the second stage of the journey, between December 22nd and the following June 21st, another balance will therefore be struck on the midway date of March 21st. Again, as on September 23rd, the noonday sun appears at 40° above the horizon at both A and B, where it is spring and autumn respectively.

You will notice (Fig. 14) that on both March 21st and

September 23rd the earth in its orbit presents a "profile" view to the sun. In relation to the sun, both hemispheres are in a neutral position, being neither tilted towards nor away from it.

Summary of the Seasons

Date	Northern Hemisphere	Southern Hemisphere
June 21st marks	summer	winter
September 23rd marks	autumn	spring
December 22nd marks	winter	summer
March 21st marks	spring	autumn

REVOLUTION AND THE "SWING" OF THE SUN

The Swing between the Tropics.—Owing to the tilt of the earth across its orbit, the noon sun is not always overhead at the

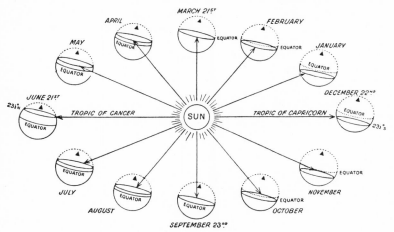

Fig. 18.—The "Swing" of the Sun

Equator. Fig. 18 shows the earth at monthly intervals along its orbit, the arrow in each case representing the sun's rays that at noon strike the earth from overhead.

During the year this overhead noon sun is seen to appear along different parallels. To make this clear, the portion of the earth lying to the north of the particular parallel that receives

43

[Kenya Information Office

A Street Scene at Noon in Mombasa

Mombasa is in Kenya, within the Tropics. What evidence is there that the noonday sun is overhead?

the vertical sun is, in the diagram, sliced off (i.e. the part bounded by dotted lines). A triangle, representing the British Isles, is shown lying within this severed part throughout the year—therefore to us in Britain an overhead sun is unknown, but is always to be found somewhere to our south. It comes nearest to us on June 21st, when it reaches parallel $23\frac{1}{2}°$ N. Note that this is the same angle as that at which the equatorial plane meets the plane of the orbit. The *Tropic of Cancer* (Greek *tropikos*=turn), as this parallel is called, thus marks the northern limit of the sun's migration. Notice how the overhead sun moves farther and farther away from us from June 21st until December 22nd, when it reaches the southern limit of its journey at parallel $23\frac{1}{2}°$ S., known as the *Tropic of Capricorn*. From December 22nd until the following June 21st it slowly creeps nearer to Britain.

On September 23rd and March 21st, the half-way stages on

its southward journey and northward journey respectively, the sun must be overhead along the Equator midway between the Tropics.

The sun therefore appears to "swing" northwards and southwards between the Tropics of Cancer and Capricorn, although, of course, it is really the earth which moves. It appears overhead at noon once a year at each Tropic, twice a year between the Tropics (at unequal intervals except at the Equator), but never to the poleward side of them.

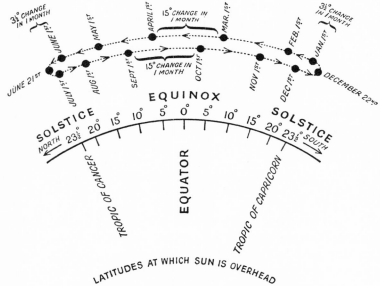

Fig. 19.—*Why the Sun "stands" at the Solstices*

Solstices and Equinoxes.—June 21st and December 22nd are termed *solstices*. June 21st is the northern summer solstice but southern winter solstice, and December 22nd is the northern winter solstice but southern summer solstice.

Solstice comes from two Latin words, *sol*=sun and *sistere*= to cause to stand. For about a month centred around each of June 21st and December 22nd the noon sun lingers, or "stands", at roughly the same angle as it slowly approaches the Tropic, reaches it, turns, and continues its "swing" in the opposite direction. Fig. 19 shows how around each solstice the sun takes a month to cover only about 3° of latitude, turn at the Tropic, and cover the same 3° on its return journey, whereas at any other time of the year a month takes it through 10° or more. Thus the sun "slows down" as it turns. You can remember that the sun "stands still" at these periods by noting the similarity in sound between solSTICE and sun "STICKS".

> *As December days grow dimmer,*
> *Once again the point is met*
> *Where the sun's receding glimmer*
> *Sinks as low as it can get.*
> *Sags the year, the solstice enters,*
> *Yet to raise the hearts of men*
> *Round one thought hope justly centres,*
> *This is where we start again.*

(From "Miscellany" Column, *Manchester Guardian*.)

March 21st and September 23rd are called *equinoxes* (Latin *aequus*=equal, *nox*=night), for on those dates over the whole world both day and night are 12 hours in length. March 21st marks the northern spring, or vernal, equinox but the autumnal equinox south of the Equator. September 23rd is the northern autumnal but the southern spring equinox.

Direction of Sunrise and Sunset.—The swing of the sun also controls the direction of sunrise and sunset, as shown in the following table.

	March 22nd to Sept. 22nd	*Sept. 24th to March 20th*
Season	⎰ Northern summer. ⎱ Southern winter.	⎰ Northern winter. ⎱ Southern summer.
Sun overhead	North of Equator.	South of Equator.
Sun rises Sun sets	North of east. North of west.	South of east. South of west.

Only on March 21st and September 23rd, when at noon it is overhead along the Equator, does the sun rise due east and set due west.

REVOLUTION AND THE LENGTH OF DAYLIGHT

THE following chart gives the Greenwich time, to the nearest quarter-hour, of sunrise and sunset at the solstices and equinoxes for Cape Wrath (North Scotland) and Penzance (Cornwall).

	Cape Wrath (58½° N., 5° W.)			*Penzance* (50° N., 5¼° W.)		
	Sunrise	*Sunset*	*Daylight*	*Sunrise*	*Sunset*	*Daylight*
June 21st	3.15 a.m.	9.30 p.m.	18¼ hrs.	4.15 a.m.	8.30 p.m.	16¼ hrs.
Sept. 23rd	6 15 a.m.	6.15 p.m.	12 hrs.	6.15 a.m.	6 15 p m.	12 hrs.
Dec. 22nd	9.15 a.m.	3.30 p.m.	6¼ hrs.	8.15 a.m.	4.30 p.m.	8¼ hrs.
Mar. 21st	6.30 a.m.	6.30 p.m.	12 hrs.	6.30 a.m.	6.30 p.m.	12 hrs.

From these times what do you learn about the duration of daylight in the extreme north of Great Britain compared with that in the extreme south (*a*) in summer, (*b*) in winter, and (*c*) in spring and autumn? Clearly, except at the equinoxes, the length of daylight and darkness changes with latitude according to the seasons.

DAY AND NIGHT IN THE NORTHERN HEMISPHERE

(*a*) *Long Days and Short Nights in Summer.*—From Fig. 20 we see that in June over half of any parallel in the northern hemisphere is in daylight, e.g. along parallel 50° N. A–B–C is longer than C–D–A.

During its daily rotation every place north of the Equator therefore passes through more daylight than darkness. The farther north a place is, the longer is its journey through light compared with that through darkness; the earlier is sunrise and the later is sunset. Eventually, when the Arctic Circle (parallel 66½° N.) is reached, rotation on June 21st takes place in complete daylight in the "land of the midnight sun".

The polar cap lying beyond the Arctic Circle is now tilted well over towards the sun, and the parallels within it (Fig. 21) are small enough to rotate more than once in sunlight, i.e. daylight lasts for over 24 hours. The nearer a parallel is to the

47

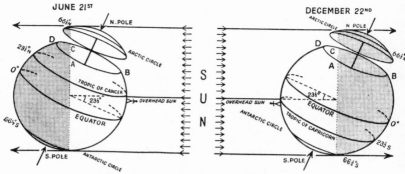

Fig. 20.—*Seasonal Changes in Length of Day and Night*

North Pole, the smaller it is and the greater is the number of sunlit rotations which it can make within the summer half-year from March 21st to September 23rd.

Thus, in Fig. 21 parallel A rotates in daylight for two months (May to July), parallel B for four months (April to August), while at the North Pole the sun never sinks below the horizon during the six summer months from March to September.

(*b*) *Short Days and Long Nights in Winter.*—When the earth takes up its December position in its orbit, bringing winter to the northern hemisphere (Fig. 20), we find that: (i) over half

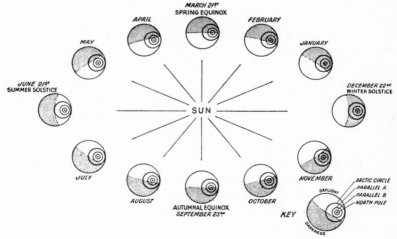

Fig. 21.—*Day and Night inside the Arctic Circle*

48

[*Popper*

The·Midnight Sun

The path of the midnight sun, as seen in summer within the Arctic Circle in Northern Canada, is shown by this photograph. On one plate a number of exposures were made at intervals from before to after midnight. Why does the path sag?

of each northern parallel lies in darkness, e.g. along parallel 50° N. A–B–C is longer than C–D–A, so that night exceeds day, and (ii) that night lengthens as we go farther north. The sun gradually rises later and sets earlier until we reach the Arctic Circle, which on December 22nd rotates in complete darkness.

The polar cap beyond the Arctic Circle is now tilted well away from the sun, and the parallels within it are small enough to make more than one complete rotation in darkness, i.e. night persists for over 24 hours. Parallel A (Fig. 21) rotates in the dark for two months (November to January), parallel B for four months (October to February), while at the North Pole the sun does not rise at all during the six winter months from September to March.

(*c*) *Equal Days and Nights in Spring and Autumn.*—Since the long days of the northern summer gradually give way to short winter days and vice versa, it follows that at the two half-way stages on September 23rd and March 21st, when neither hemisphere faces the sun, a balance must be reached. Consequently, on these two dates day equals night in length (the equinoxes).

DAY AND NIGHT IN THE SOUTHERN HEMISPHERE

FROM Fig. 20 can be worked out a seasonal pattern of day and night for the southern hemisphere also. It resembles that found north of the Equator, although we must not forget that the conditions which prevail in the northern hemisphere in June (summer) and December (winter) will apply in southern latitudes to December and June respectively. The Antarctic Circle (66½° S.), the southern counterpart of the Arctic Circle, basks in its midnight sun on December 22nd and receives its 24 hours of complete night on June 21st. At the South Pole the sun never sets from September to March and never rises from March to September.

As in the northern hemisphere, March 21st and September 23rd are the equinoctial "balancing" periods, when everywhere daylight equals darkness in length.

DAY AND NIGHT AT THE EQUATOR

SINCE at all times but the equinoxes conditions north of the Equator fade into the reverse south of it, a balance between the two differing hemispheres must always be struck along the Equator, where day equals night throughout the whole year.

EXERCISES

1. What climatic changes would take place in polar regions if the earth's axis tilted over until it met the plane of the orbit at an angle of about 45° ?

2. Keep a record, making entries at regular weekly intervals for as much of the year as possible, of (i) the direction of the sun from your school at the following set times: 9 a.m., 12 noon (make allowance for any clock changes in "summer time"), and 3 p.m., and (ii) the sun-angle above the horizon at noon. What do your observations tell you about the changes in the sun's daily path across the sky throughout the year?

3.

Place	June 21st		December 22nd	
	Sunrise	Sunset	Sunrise	Sunset
Wick (58½° N., 3° W.)	3.10 a.m.	9.20 p.m.	9.10 a.m.	3.20 p.m.
Liverpool (53½° N., 3° W.)	3.50 a.m.	8.40 p.m.	8.30 a.m.	4.00 p.m.
Dublin (53½° N., 6° W.)	4.00 a.m.	8.50 p.m.	8.40 a.m.	4.10 p.m.

(*a*) Why in June is sunrise earliest and sunset latest at Wick?

(*b*) Why in December is sunrise latest and sunset earliest at Wick?

(*c*) Why, although both are on the same parallel, does Dublin lag behind Liverpool in sunrise and sunset in both June and December?

CHAPTER V

Climate and Weather

HOW CLIMATE DIFFERS FROM WEATHER

CLIMATE is not necessarily the same thing as average weather. In fact, to apply the word *average* to the weather in many regions is impossible, as it is so changeable. For instance, it is ridiculous to average yesterday's steady downpour and to-day's bright sunshine by describing both days as "showery with bright intervals".

If, however, it rains on two or three days out of every four or five in a particular season, we can call the climate for that season wet. We are likewise justified in describing a winter as mild, if, as in Britain, the total length of the isolated spells of really cold weather does not exceed, at the most, four or five weeks. Similarly, if occasional heat waves are few and far between, a summer may be called warm. In discussing climate our attention is thus focused upon the normal state of affairs, although we cannot dismiss as unimportant short spells of unseasonable weather which sometimes occur. By climate is meant the conditions of temperature, rainfall, cloudiness, winds, humidity (i.e. the amount of moisture present in the air), etc., to be expected on the whole over a long period such as a season, whereas weather means the combination of these atmospheric conditions existing at the moment.

Weather often changes from day to day and even from hour to hour, but the climate of a region usually repeats itself from year to year, although abnormal years may prove exceptions to the rule. One such year, 1947, brought to Britain an unusually cold and snowy winter, during which rivers and even coastal waters froze, succeeded by an exceptionally hot, dry, and sunny summer.

Britain is a country where weather and climate fail to agree. A British winter is climatically classed as "mild", but usually

[*Picture Post*

An Abnormal Winter in Britain

British weather is noted for its variety. Winters are usually mild, but occasionally abnormal weather occurs, as shown by this view of the frozen Thames in 1894.

consists of cool, wet days followed by cold, dry periods. In February, 1948, people in Kent were snowbound one week and sun-bathing the next. Similarly, the cool, rainy days of the "warm" British summer are interrupted by hot, sunny intervals. It is therefore not surprising that the Englishman can rely upon the weather as a never-failing topic of conversation, and that British newspapers fully recognise its news-value.

In many lands, on the other hand, weather and climate are practically alike. An Arab who remarked, as the people of Britain so often do, "Hot to-day, isn't it," or "Nice weather we're having!" would probably have his sanity questioned, for such statements are quite unnecessary in the continuous heat wave of desert Arabia. In some climates a cricket match could be arranged for months ahead with the certainty that rain would not stop play. Again, so punctual is the afternoon thunderstorm in certain equatorial lands that an appointment can be made "after the storm" as we say "after tea". In India famine threatens should summer rains fail to pour from leaden skies, while Russians take for granted their icy winters and scorching summers. In such cases as these there is, of course, some value in defining climate as "average weather".

MEANS OF TEMPERATURE AND PRECIPITATION

WE cannot always rely upon our feelings as a guide to climatic conditions, for our senses sometimes give us a false idea of the actual facts. For instance, cool but damp weather often causes us to shiver more than we do when it is really cold but dry. Similarly, sweltering in the oppressive heat of a hot, moist day, we tend to imagine that the temperature is higher than it actually is and we feel less comfortable than we do on an extremely hot but dry day. For these reasons the dry, invigorating cold of a Canadian winter is sometimes to be preferred to the raw chilliness of a British winter day, while dry desert heat is less trying than the damp, "sticky" heat of equatorial regions. Scientific descriptions of climates are, however, based upon figures of temperature, rainfall, etc., about which there can be no uncertainty or argument.

In any discussion about a climate the two chief items to be considered are temperature and precipitation, a word meaning moisture in all its various visible forms—rain, dew, hail, fog, and snow. To choose at random the weather of any particular day as typical of the climate of a whole season would be absurd. That day might be unusually warm or cold, wet or dry. Accounts of seasonal changes in climate are therefore based upon monthly *means*, or average figures of temperature and precipitation. These are acquired from a network of weather or meteorological stations scattered over the civilised world, and also from weather ships stationed on the oceans.

HOW MEANS ARE OBTAINED

SUPPOSE that the mean January temperature for a certain place is required. Mounted in a white wooden box known as a Stevenson screen, specially constructed thermometers record the maximum and minimum temperatures reached during January 1st. These figures are read at a fixed time, e.g. 9 a.m., on January 2nd. The two thermometers are then set to the actual temperature at 9 a.m. and the process is repeated, the maximum and minimum temperatures for January 2nd being observed at 9 a.m. on January 3rd, and so on throughout the month. The average of the maximum and minimum temperatures for each day is taken as the mean temperature for that day, although sometimes the mean is calculated in other ways. At the end of January the 31 separate daily means are totalled and divided by 31 to give the mean temperature for the whole month. This particular January, however, might be much milder or colder than usual. Therefore, to give a true picture of a normal January, the process is repeated over a period of at least 35 years. The total of these varying means is divided by the number of years to obtain the final long-period mean for the month. In spite of a number of unseasonable exceptions either well above or below it, the average January temperature for most years will approach this value.

The mean January rainfall figure is obtained by measuring in a rain-gauge the daily rainfall (if any). The total amount for the whole month is then calculated. In this case, however, the monthly figure is not divided by 31, since a daily average would serve no useful purpose, and would probably be both misleading and meaningless. This process is repeated for many years and the various January totals are added together and averaged to furnish a long-period mean. This gives a satisfactory idea of the rainfall to be expected in most Januaries, although, as with temperature, certain unseasonable exceptions will occur. Such exceptional weather cannot be ignored, for it often plays a great part in human affairs. In spring sharp night frosts, not suspected from a study of mean temperatures, may destroy all hope of a good fruit crop in Kent, while abnormally heavy rain in late summer may ruin an East Anglian wheat harvest.

Air pressure, amount of sunshine, and other items of weather can also be averaged.

WORD-SCALES AND THE DESCRIPTION OF CLIMATES

WE can convert the temperature and precipitation figures for any place into a written account of its climate by using word-scales, in which particular descriptions correspond to particular temperatures or annual amounts of rainfall. In later chapters the climates of the world are described according to the following word-scales for (*a*) mean temperature, (*b*) annual range of temperature, i.e. the difference between the highest and lowest mean monthly temperatures, and (c) annual amount of precipitation, i.e. how much moisture falls in the year:

Word-scales

Mean Temperature

° F.		° C.	Climatic Description
Over 86	or	30	Very hot.
68 to 86	,,	20 to 30	Hot.
50 to 68	,,	10 to 20	Warm.
32 to 50	,,	0 to 10	Cool (or mild).
14 to 32	,,	− 10 to 0	Cold.
Below 14	,,	Below − 10	Very cold.

Note how the intervals of 18° Fahrenheit correspond to increases of 10° on the Centigrade scale.

The temperature of the air is its degree of hotness, and this can be described only by means of a scale of numbers on the thermometer.

It is therefore incorrect to speak of *hot*, *warm*, or *cold* temperatures, for applied to a number such adjectives lose all meaning.

They can certainly be used to describe the air itself, but temperatures must be termed *high*, *moderate*, or *low*.

Annual Range of Temperature		Annual Amount of Precipitation	
° F.	Description	Ins. of Rain	Description
Over 60	Very great (or extreme).	Over 60	Very heavy.
40 to 60	Great.	40 to 60	Heavy.
20 to 40	Moderate.	20 to 40	Moderate.
10 to 20	Small.	10 to 20	Light.
Under 10	Very small.	Under 10	Very light.

Note how, although the same sets of figures appear on these last two word-scales, the descriptions differ. We speak of a "heavy" or "light" rainfall, not of a "great" or "small" one.

In describing a climate, no really important aspects of it will be omitted if the following scheme is always adopted:

A. *Temperature.*

    1. Describe the *Summer Season*, e.g. very hot, warm, etc.

    2. Describe the *Winter Season*, e.g. very cold, mild, etc.

    3. Describe the *Annual Range of Temperature*, e.g. very great, moderate, etc.

B. *Precipitation.*

Describe this under the headings:

    1. *Annual Amount*, e.g. very light, heavy, etc.

    2. *Seasonal Distribution.* There is no word-scale to show in what seasons rain falls. The total rainfall for the summer half-year should be compared with that for the winter half-year. The distribution can then be described as follows: "summer drought and winter rain", "rain at all seasons", "a summer maximum" (i.e. well over 60 per cent. of the annual total falls in the summer half-year), etc.

    3. *Form and Cause.* Precipitation takes different forms. Wherever winter temperatures fall well below freezing-point it falls as snow. Morning dews are characteristic of hot deserts. Rainfall, as we shall learn in later pages, may be caused in three ways: by rising currents of hot air; by winds blowing over mountains; and by the meeting of cold and warm air currents.

CLIMATE AND MAN

MAN's activities are largely controlled by climate. The food he eats (at least, that part of it which is not imported from other regions), the clothes he wears, the shape and structure of his dwelling and the material of which it is built, the width of the streets in his towns, and even his personal habits and physical appearance—all of these can be, directly or indirectly, traced to the climate of his home region.

The food crops that man grows differ with differing climates. The Chinese coolie and Indian ryot, or peasant, eat rice; the equatorial Negro depends upon yams, cassava, and bananas;

Climate and Architecture

[*Allan Cash*

The overhanging eaves of these Swiss chalets keep doorways and windows free from snow. Timber for building material comes from coniferous forests, which often grow in regions where winters are cold.

and to the South Sea Islander the coconut palm is invaluable. Similarly, the animals which feed and clothe man, which provide skins for his tents, and which transport him and his goods vary in different climatic regions. The Lapp relies on his reindeer, the Tibetan on his yak, and the Peruvian on his llama; the Arab depends on the mare, goat, and camel; the herdsman of African savannas counts his wealth in cattle, and the Asian Khirgiz in goats, sheep, and horses. In arctic tundras the Eskimo catches the seal, caribou, and fish, while in Australian deserts the aborigine eats whatever living creature he can find.

While the Laplander wraps himself in warm skins and the Bushman of the Kalahari Desert is practically naked, the Russian changes his costume according to the season—furs in winter snows and loose-fitting light garments in summer heat waves.

The Eskimo shelters from winter blizzards in an igloo of

57

snow or of earth and stone; the Khirgiz in the Asian steppes lives in a felt yurt; the Bedouin sets up in the desert his goatskin tent, and the Australian "Blackfellow" takes refuge in his mia-mia, a flimsy bark hut.

Desert roof-tops are flat, as little rain is to be shed from them; some people here keep cool in underground cellars in the very hot weather, while narrow streets shut out a scorching sun. By contrast, wide city streets in cooler lands welcome in sunlight and warmth; here rain is shed from sloping roofs, which in countries like Switzerland project well beyond the walls so that heavy winter snows slide clear of doors and windows.

Even the physical appearance of man is partly to be explained by climate. Black-skinned Negroes are fitted for life in tropical lands, white-skinned races are found where cooler conditions prevail, while the yellow-skinned Mongolian peoples originated from dry regions with severe winters. The hawk-like nose of steppe-land and desert nomads, the broad nose of tropical peoples, and the expanded chest and deeply-breathing lungs of Andean highlanders—all have their origin in climate.

Climate also affects the physical appearance of the earth. It helps to make the scenery of the "stage" on which man, the actor, plays his part. Snow feeds the glaciers and rain the rivers, which, biting into the earth's surface, provide valley contrasts to frost-shattered mountain tops. Wind piles up sand into dunes. Moreover, the earth's carpet of vegetation changes its design with changing climates. The waving grassland of treeless prairie fades polewards through coniferous forest to arctic tundra, while desert wastes merge through tropical savanna parkland into luxuriant equatorial forest.

CLIMATE AND CIVILISATION

SOME early civilisations arose in countries where the climate was such that man could grow food crops only by irrigating the land, and in puzzling about the best way to use the precious water he exercised his brain, and so emerged from a savage state.

The most highly developed peoples of the world live where the climate is neither too hot nor too cold. In equatorial forests the pygmy, living on what he can hunt or gather from the trees and needing little clothing or shelter, finds it too hot to make strenuous efforts that develop the brain. In Arctic regions the Eskimo is too busily occupied in finding enough food to eat to spare time for the cultivation of his mind. In temperate lands between these two extremes man must work in order to satisfy his needs, yet the invigorating climate allows him sufficient leisure to direct his mental powers along lines which lead to a high level of civilisation.

In the course of time climate slowly changes. In some lands where it once favoured the growth of knowledge the climate has changed for the worse, and traces of decayed civilisations are found buried in dense tropical forests or beneath desert sands. There is some evidence, such as the retreat of arctic glaciers, that the climate of temperate and polar regions is at present becoming warmer.

Climate and weather even help to turn the course of history. Some people believe that many of the attacks that in times past were made upon settled agriculturists, i.e. crop-growers, by nomadic pastoralists, or wandering herdsmen, had their origin in climatic changes which, leading to a food shortage, drove these invaders from their homelands. Napoleon could with justification partly blame the weather for his downfall. His Grand Army vanished in the snows of an unexpectedly early Russian winter, and the French novelist Victor Hugo relates how rain on the eve of Waterloo made the ground soft and turned the tide of battle in Wellington's favour.

EXERCISES

1. February fill-dike; March comes in like a lion, and goes out like a lamb; April showers; Maytime; flaming June; November fogs. What do these descriptions tell of Britain's climate?

2. Describe the climate shown by the following mean figures for a place in the southern hemisphere:

	J.	F.	M.	A.	M.	J.	J.	A.	S.	O.	N.	D.
Temperature (° F.)	74	74	71	67	61	57	55	56	58	61	66	71
Precipitation (ins.)	0·3	0·5	0·7	1·6	4·9	6·9	6·5	5·7	3·3	2·1	0·8	0·6

3. Find out what part the weather played in the Battle of Agincourt (1415); the defeat of the Spanish Armada (1588); the Great Fire of London (1666); the "Protestant Wind" in the Revolution of 1688; the Black Hole of Calcutta (1756); the 1854–55 winter in the Crimean War; and the evacuation of the British Army from Dunkirk (1940).

CHAPTER VI

Temperature and its Chief Controls

TEMPERATURE AND HEIGHT ABOVE SEA-LEVEL (ALTITUDE)

WHY does height coat an aeroplane with ice, and why are lofty mountains snow-capped? Surely temperature should not decrease but increase as we get nearer to the sun? To solve this puzzle we must find out how air is warmed by the sun's rays.

Music broadcast from a wireless station passes through the air as inaudible waves of energy, and before you can hear it you need a receiving set, which traps these waves and transforms them back into sound. Similarly, the sun's heat streams out

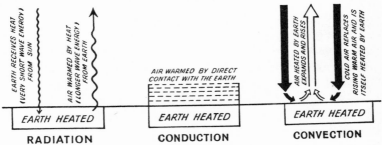

Fig. 22.—*How Air is Heated*

through space as energy in a form which penetrates most of the earth's atmosphere without heating it, although the moist, dust-laden lower layers are warmed to some extent. The land (or sea), however, acts as a "receiver", first turning this solar energy into heat and then, by *radiation*, returning it to the atmosphere in such a way that the air can be warmed. Air that is in direct contact with the heated surface of the earth also gains heat by *conduction*. A third way in which air may be warmed is by *convection*. The air heated by conduction expands and rises in *convection* currents. It is replaced by colder air, which is in its turn heated. In these three ways heat is eventually distributed throughout the lower layers of the earth's atmosphere. The transfer of heat from earth to air by radiation, conduction, and convection is illustrated in Fig. 22.

Since air is warmed by the earth rather than directly by the sun, the farther we rise the colder does the air become, just as the farther we retreat from a wireless set the fainter grows the sound. The rate of loss of heat with gain in height is called the lapse-rate of temperature (lapse = fall). The average lapse-rate is 1° F. for every 300 feet rise.

Starting in a heat wave of 80° F., we could therefore climb a mountain 18,000 feet high (about 3½ miles) on the Equator and at the summit find ourselves in air at a temperature of $80° - \dfrac{18,000°}{300} = 20°$, or 12° below freezing-point. To experience so great a fall at sea-level we should have to journey several thousands of miles towards the Poles. In actual fact, the drop in temperature would probably be less than 60° F., for in rainy regions like equatorial lands the lapse-rate is often less than the average one of 1° F. per 300 feet rise.

In tropical lowlands white men find life almost unbearable, but they can live in comfort in the cooler highlands, e.g. British colonists who have settled in the high plateau of East Africa.

Like the skins of an onion, the atmosphere consists of different layers. Beyond the ground-layer, or *troposphere*, temperature ceases to fall and remains steady throughout the next layer, a calm and cloudless zone called the *stratosphere*. In one of the higher layers temperature actually increases until the weather is as warm as it is on the earth's surface.

TEMPERATURE AND DISTANCE FROM THE EQUATOR (LATITUDE)
BETWEEN the Tropics the mid-day sun is always high in the sky and twice a year is overhead. Consequently, tropical lands enjoy a perpetual heat wave, and seasons are distinguished from one another less by changes in temperature than by variations in rainfall. "Wet" and "dry" seasons here take the place of the "summer" and "winter" of temperate lands, although near the Equator the so-called "dry" season is very short and is

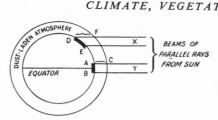

Fig. 23.—How Latitude affects Temperature

Rays X serve area DE
Rays Y serve area AB
DF — path for rays X
AC — path for rays Y

by no means a drought.

Polewards from the Tropics, however, temperature decreases for three reasons:

(*a*) Increasing latitude brings about a decrease in the noon-day sun-angle (Figs. 15 and 23).

(*b*) As the Poles are approached the sun's rays take an increasingly long path through the atmosphere (Fig. 23). Dust particles and water-drops in the air rob the rays of some of their heat, and the longer their aerial journey is, the greater is the amount of heat thus "stolen" before the earth itself benefits from them.

(*c*) Winter becomes particularly severe in arctic and antarctic lands, where daylight is so short that the sun gets practically no chance to warm the earth, and where in mid-winter it does not rise at all. In summer, of course, the converse holds good, and long hours of daylight help to make up for the low mid-day sun-angle.

TEMPERATURE AND DISTANCE FROM THE OCEANS

A WORLD consisting entirely of either land or water would gradually grow colder in a regular fashion from tropical heat to polar snows. But for the cooling influence of mountains, temperatures for a given season would be practically constant along any particular parallel. Such a regular pattern of falling temperatures from Equator to Poles is, however, upset by the jig-saw arrangement of continents and oceans.

Substances differ greatly in the rate at which they absorb and lose heat. Water gains warmth very slowly and is in no hurry to part with it. With land, however, it is a case of "quickly come, quickly go". So rapidly is it heated that in desert battles troops have sometimes preferred to stand up and risk death than to take cover on the burning sand. In winter, and after sunset, the land loses heat as fast as it absorbed it when the sun

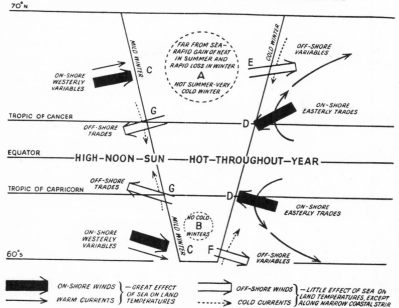

Fig. 24.—*The Effect upon Temperature of Prevailing Winds, Currents, and Distance from the Sea*

was high in the sky, particularly when the sky is cloudless.

These natural laws, of course, apply to continents and oceans. Fig. 24 represents an imaginary continent extending polewards from the Equator. Like the real continents, it reaches 70° N. but goes no farther than 60° S., for in the southern hemisphere only Tasmania, New Zealand, and the tip of South America continue beyond 40° S. Our imaginary land-mass also tapers southwards, just as bulky North America and Eurasia contrast with the "tail-ends" of the southern continents.

As atmospheric temperature is largely controlled by the surface upon which the air rests, maritime (or oceanic) air will everywhere show a smaller range of temperature than continental air. The former stays pleasantly warm in summer as the oceans gradually absorb heat, and remains mild in winter when the water retains much of this slowly acquired warmth. The temperature of continental air, however, violently sweeps up in summer and down in winter, as heat is first greedily taken in by the land and then rapidly radiated away into space. In the perpetual summer of equatorial latitudes the difference between maritime and continental air is not important (Fig. 24), but polewards from the Equator the contrast between them increases as the rhythm of summer and winter grows stronger.

Maritime and continental air differ most in temperature wherever a continent (*a*) extends far polewards and (*b*) is very wide, with its heart removed thousands of miles from the equalising influence of the oceans. It so happens that conditions (*a*) and (*b*) coincide in the great northern land-masses of North America and Eurasia. On the other hand, the southern continents (*a*) do not reach far beyond 40° S. (Africa extends only to 35° S.), and (*b*) taper polewards, so that they cannot develop that great annual range of temperature which is so characteristic of the interior of their wider northern counterparts (contrast A and B, Fig. 24).

THE CONTINENTAL CLIMATES OF CONTINENTAL INTERIORS

THE interior regions of huge land-masses are quite beyond the influence of the far-off oceans, and the climate here is for obvious reasons called *continental*. The most remarkable examples of this climate are to be found in the wide northern continents (Fig. 24, Region A). Here a bitterly cold and long winter follows on the heels of a short but hot summer. In North America, Winnipeg's mean January temperature of −4° F. has by July soared to 66° F. (range 70°). The "Cold Pole" is in Siberia at Verkhoyansk, which holds the world's record for cold winters; its mean July temperature of 60° F. sinks to nearly −60° F. in January (range 120°), while in February 1892 the thermometer registered 122° of frost (−90° F.).

A sergeant in Napoleon's army, describing the famous retreat from Moscow, writes: "This terrible cold was more than I had ever felt before. I was almost fainting, and we seemed to walk through an atmosphere of ice. . . . I could hardly breathe: my nose felt frozen; my lips were glued together; my eyes streamed, dazzled by the snow. I was forced to stop and cover my face with my fur collar to melt the ice. . . . I saw an old trooper of the Imperial Guard, his moustaches and beard covered with

[S.C.R.

A Continental Winter

This scene in Moscow illustrates the very cold winters of the continental interiors of Eurasia and North America, far from the warming influence of the ocean.

icicles . . . he related how his fingers had frozen before reaching Smolensk. During that night all his fingers fell off one after the other . . ." (Memoirs of Sergeant Bourgogne).

In such low temperatures fur clothing is essential. Additional "storm" windows and doors are fitted on houses, motor cars are equipped with heating devices, and in the more modern towns central heating apparatus spreads warmth throughout the buildings. So cold is it that the water-vapour in air breathed out through one's nose is immediately condensed into tiny ice crystals. The coldest weather experienced in Britain occurs, usually in January or February, whenever this icy air streams westwards from Russia to envelop our country, giving us a brief taste of a continental winter.

The continental summer, however, is a season for sun-bathing. A hot sun blazes down from an almost cloudless sky, marred only by occasional thunder-clouds from which pour heavy but short-lived showers. To keep cool, the Russian shaves his head and wears a loose, white jacket; doors and

[*E.N.A.*

A Continental Summer
What evidence of heat is given by this picture of Moscow in summer? The oceans cool the land in summer, but continental interiors are too far inland to benefit from their cooling influence.

windows are flung wide open and are covered only by netting to shut out mosquitoes and flies.

In the southern hemisphere, where for every thousand square miles of land there are tens of thousands of square miles of water, and where the land-masses taper, the continental winter is not cold and the annual range of temperature is moderate rather than great (Fig. 24, Region B).

The *temperate* continental type of climate as described here should not be confused with the *tropical* continental type which forms the subject of a later chapter. Both climates can be called continental because the lands that experience them are well inland, far from the ocean's influence. Situated within the Tropics, however, tropical continental regions escape the bitter winters of temperate continental lands, for at all seasons the noon sun stands high in the sky and "winter" is therefore scarcely less hot than summer.

THE MARITIME CLIMATES OF REGIONS WITH ON-SHORE WINDS

IN certain regions on the margins of continents the annual range of temperature is much smaller than is warranted by their latitudes. Such regions are indebted to the oceans for the mildness of their climates, which are therefore described as *maritime*.

By its own unaided efforts the sea can influence only a very narrow coastal strip. To give its maximum benefit to the land, therefore, maritime air must somehow be brought well inland. Such a transference of air from sea to land is possible only if the prevailing winds, i.e. those which normally blow, are on-shore. Not all *marginal* lands, as those which face the ocean are called, have a maritime climate, for in some cases the prevailing winds are off-shore. Whenever this happens the oceans might just as well not exist so far as any control they have over the land is concerned, for off-shore winds carry the sea's moderating influence away from continents. Even with the aid of on-shore winds maritime influences will not penetrate far inland if high north-south mountains such as the Andes impose a barrier to the progress of the winds. Thus nearness to the sea does not necessarily guarantee a maritime climate.

In each hemisphere maritime influences are carried ashore by two great sets of prevailing winds: (*a*) in temperate latitudes the *variables* blow from westerly directions, and (*b*) in tropical and sub-tropical regions (as lands just outside of the Tropics are called) the steady *trade winds* blow from the east. It so happens that wherever all these winds blow on-shore they first pass over warm ocean currents (Fig. 24). The winds drive the currents along in their own direction and winds and currents, the one above the other, move along together towards the continents. Constantly warmed by the currents, the winds are remarkably mild by the time they reach the land, which benefits accordingly. In each hemisphere, therefore, two separate maritime climates correspond to the two different wind systems, as follows:

(*a*) *The Temperate Maritime Climate.*—This climate is experienced along those western margins of continents which receive on-shore westerly variable winds (Region C, Fig. 24). Other names given to it are *western cool temperate, oceanic,* or *British,* for maritime influence is particularly strong over the British Isles. Temperate maritime lands enjoy winters which, judged by the average for their latitudes, are much milder than they should be. Especially is this so in North-western Europe, where the variables bring ashore warmth from the North Atlantic Drift, a remarkably warm ocean current; moreover, here there are few north-south mountain ranges to interrupt the progress inland of these mild winds.

Unlike the hot, cloudless summers of the continental climate, temperate maritime summers are cool, cloudy, and changeable, with occasional hot spells. Contrast, for instance, continental Winnipeg's annual range of temperature of 70° (from −4° F. in January to 66° F. in July) with Falmouth's range of 17° (from 43° F. to 60° F.). Both places are on the same parallel of 50° N., but Falmouth has a temperate maritime climate.

(b) *The Tropical Maritime Climate.*—This climate occurs along those eastern margins of continents towards which trade winds blow from the oceans (Region D, Fig. 24). Throughout the year temperatures here, as befits a tropical position, are considerably higher than in temperate maritime regions. The annual range is small, e.g. 11° at Rio de Janeiro (February 79° F., July 68° F.).

THE CLIMATE OF REGIONS WITH OFF-SHORE WINDS

(a) *Eastern Cool Temperate Climate.*—Portland (Oregon) is on the west of the United States, about 100 miles from the Pacific Ocean. Portland (Maine) is an Atlantic sea-port on the east of the same country. Both cities are thus on the margins of North America, and both are practically on the same parallel. It would therefore be reasonable to suppose that both have the same kind of climate. This belief seems to be confirmed by their summer temperatures of 67° F. in July in Oregon and 68° F. in Maine.

Winter tells a different story. Pacific Portland enjoys the mild winter of a temperate maritime climate, with a mean January temperature of 39° F. At its Atlantic namesake, however, the temperature falls to 22° F., or 10° of frost; obviously the city, although it is coastal, receives no warming maritime influence. It has an *eastern cool temperate* climate, sometimes called the *Laurentian* climate because it is characteristic of the St. Lawrence Valley of Eastern Canada (Region E, Fig. 24).

The reasons for this severe eastern winter are again to be sought in the direction of prevailing winds and ocean currents. Fig. 24 shows that here (i) the prevailing westerly winds are off-shore and bring with them the bitingly cold weather of the

[U.S. Information Service

Winter Snow in North-eastern U.S.A.

Winters are cold in the eastern cool temperate margins of North America and Asia. Why are such severe winters unknown in the southern hemisphere?

continental heart; (ii) the warm ocean current, which for temperate maritime margins was on-shore, is now moving away from the land; and (iii) flowing along the shore and to some extent affecting coastal temperatures is a cold current, bringing with it icebergs from arctic seas. Thus, while New York's summer heat waves rival those of India, the city often suffers from winter blizzards, as on Boxing Day 1947, when 26 inches of snow fell in 16 hours. On one occasion water from a fireman's hose-pipe froze before it reached the flames. Yet along the Pacific Coast in the same latitudes snow is practically unknown.

South of the Equator the absence of wide land-masses in these latitudes prevents the development of this kind of climate. The nearest approaches to it occur in Patagonia in Southern Argentina and on the east side of the South Island of New Zealand (Region F, Fig. 24). Here, however, since there are no wide continental interiors with cold out-blowing winds, winters are not severe. For instance, Dunedin (New Zealand) has

69

a range of only 16° F., from 58° F. in January to 42° F. in July.

(*b*) *Hot Desert Climate.*—Just as the on-shore westerly variables of temperate maritime lands become off-shore over eastern cool temperate margins, so the easterly on-shore trade winds of tropical maritime lands become off-shore over the opposite western margins. These off-shore trades push away from the land any control which the sea might otherwise exert over it, and only a narrow coastal strip comes under the sway of oceanic influences. The *hot desert* type of climate prevails over these western margins (Region G, Fig. 24).

TEMPERATURE AND OCEAN CURRENTS

APART from the ocean itself, currents of warm or cold water sliding over its surface often affect the temperatures of marginal regions. These currents and their influence upon climate are described in Chapter XIV.

EXERCISES
1.

Town	Feet above sea-level	Latitude	January temperature	July temperature
Singapore	10	1° N.	80° F.	81° F.
Brest (W. France)	200	48° N.	45° F.	65° F.
Quebec (E. Canada)	300	47° N.	10° F.	66° F.

Assuming a lapse-rate of temperature of 1° F. per 300 feet rise in height, calculate how high above Singapore an airman must fly in (*a*) January and (*b*) July to reach mean temperatures equal to those of Brest and Quebec. Remembering that these two towns are alike in both latitude and height above sea-level, explain why your answers for them are very different for January but similar for July.

2. Explain what lies behind the saying that "in Central Canada butter is sold by the pint in summer and milk by the pound in winter".

3. Certain rivers are given below in their correct order from west to east across Eurasia. In brackets is shown the average number of days per year for which each is frozen at about parallel 51° N.—Thames (0), Rhine (20), Vistula (70), Dneiper (100), Don (110), Volga (120), Irtish (140), Amur (170). Give a full explanation of what these figures prove about winter temperatures as one penetrates Eurasia eastwards from the Atlantic Ocean.

CHAPTER VII

Temperature and its Minor Controls

TEMPERATURE AND CLOUDINESS

Cloudless Days and Nights.—Why are both the hottest and coldest days heralded by clear skies? We already know that the sun radiates heat through the atmosphere to the earth, which in turn re-radiates it back to the air. During summer days, with a sun high in the sky and many hours of daylight, the earth receives more solar heat than it can send out. When skies are

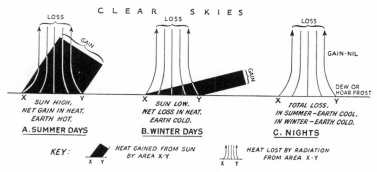

Fig. 25.—*The Effect of Clear Skies on Temperature*

clear nothing can prevent this gain of heat from proceeding at its maximum rate and so causing a heat wave (Fig. 25A). In winter, on the other hand, the sun is much lower in the sky and the hours of daylight are fewer. The earth now loses into space far more heat than it can replace from the low, short-lived sun. Thus it rapidly spends the balance of heat which it had stored up during summer days of plenty. Whenever winter skies are clear this loss of heat is very rapid indeed; the cooling earth chills the air above it and days are frosty (Fig. 25B).

Why are winter frosts sharpest when the stars twinkle brightly,

71

and why do the heaviest dews follow the clearest nights? During the hours of darkness no solar heat is received, but radiation from earth to sky continues. On a clear, starry night this heat escapes freely and unchecked, and the temperature falls rapidly. The air in contact with the chilling earth is cooled to such an extent that its invisible water-vapour is condensed into dew-drops. Should the temperature fall below freezing-point, as it often does in winter, the water-vapour is condensed as hoar frost (Fig. 25c).

Cloudy Days and Nights.—What makes cloudy days cool in summer but mild in winter? Clouds may be compared to a

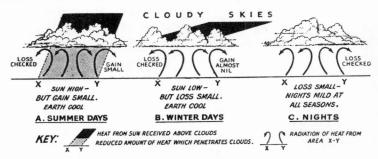

Fig. 26.—*The Effect of Cloudy Skies on Temperature*

huge blanket hovering above the earth. They prevent solar heat from reaching the surface, so that cloudy summer days are cool —you have all felt the sudden chilliness caused when a passing cloud masks the sun (Fig. 26A). In winter, however, the cloud blanket checks the radiation of heat from the earth and, by trapping this heat and preventing its escape, keeps us warmer than we should otherwise be (Fig. 26B). The same blanketing effect applies to cloudy nights, which are always mild and muggy in contrast to cool or frosty clear, star-lit nights (Fig. 26c).

Clear and Cloudy Regions.—Some regions are more cloudy than others. In equatorial lands huge thunder-clouds tower upwards during the hot afternoon, whereas desert skies are clear. Consequently, although day-time temperatures in equatorial latitudes rarely reach 100° F. they may in the desert soar above

[*Swiss National Tourist Office*

The Effect of Aspect in an East-to-West Valley in Switzerland

In this view of the Lötschenthal in Switzerland what evidence is there that the slopes shown on the left of the picture face southwards whereas those on the right face northwards?

130° F. in summer (Tripoli, in North Africa, holds the world record maximum temperature of 136° F.). Starry desert nights, however, are as cold as the days are hot; in summer they are cool, while in winter fires are lit after sunset to combat the bitter cold.

Following these cold nights, early morning dews are particularly heavy in deserts, for even in arid lands the air contains some moisture. The equipment of French Foreign Legionnaires includes a heavy blanket—a nuisance during the hot marches of the day, but an absolute necessity during the icy night. On the other hand, cloudy equatorial nights remain warm and muggy, and the temperature rarely falls to 60° F.

Temperate maritime lands are also noted for cloudy skies— yet another reason for their comparatively cool summers and mild winters. Continental regions, however, lie so far from the sea, the chief source of cloud and rain, that they are cloudless but for an occasional summer thunderstorm. Here the rapid gain of heat in summer and its equally rapid loss in winter are greatly aided by this absence of a cloud blanket.

Monsoon lands such as India provide another good example of the effect of clouds upon temperatures. Here the hottest weather occurs in the dry, cloudless spring, for in summer the temperature falls slightly when steady rains stream from overcast skies. The hot, moist air of equatorial and monsoon regions is, however, far more trying for human comfort than the even hotter but dry air of deserts.

TEMPERATURE AND ASPECT

WHEN the sun breaks through after a heavy snowfall, snow often disappears faster on one side of a roof than on the other.

Fig. 27.—N. to S. Section across an Alpine Ridge, showing the Effect of Aspect on Temperature and the Snow-line

On the warmer side, facing the sun, it thaws readily, whereas it lingers in the shade of the cooler opposite side.

In much the same way, a hillside facing the sun is warmer than ground which slopes away from it. For instance, on the sunny southern slopes of east-west Alpine ridges in Switzerland, the *snow-line*, i.e. the level above which snow remains unmelted, is often at least 500 feet higher than it is on the shady northern slopes (Fig. 27).

In our own country we all know how a room with a warm southerly aspect is preferable to one with a cool northerly outlook. In similar southern latitudes the converse would hold good.

TEMPERATURE AND LOCAL WINDS

APART from great planetary winds-systems such as the trades and variables, there are certain winds, warm or cold, pleasant or unpleasant, which operate only in special localities but which greatly affect temperatures in these areas.

Among the more important local winds are the following:

Name		Area
Cold winds	Bora	N.E. Adriatic Coast.
	Mistral	S. France.
	Pampero	Argentina.
	Buran	Siberia.
	Norther	Southern U.S.A.
	Southerly Burster	New South Wales.
Warm winds	Khamsin	Egypt.
	Brickfielder	S.E. Australia.
	Santa Ana	S. California.
	Harmattan	From Sahara to Guinea Coast.
	Sirocco	From Sahara to S. Europe.
	Nor' Wester	Canterbury Plains (N.Z.).

Protective wind-breaks of hedges or trees are planted to shelter crops from the most violent and destructive of these winds, e.g. the pampero and the mistral. The latter has even been known to overturn trains.

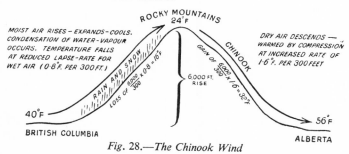

Fig. 28.—The Chinook Wind

One of the best-known local winds is the Chinook. This warm wind descends from the Rockies to the prairies of Alberta and the west of the United States. It originates as a westerly wind which is forced to rise by the western slopes of the Rockies, e.g. in British Columbia. When air rises it expands and cools, until eventually its invisible water-vapour is condensed into raindrops or snowflakes. Rising air, if accompanied by this

75

condensation of water-vapour, loses heat at less than the average lapse-rate of 1° F. per 300 feet—namely at 0·8° F. per 300 feet.

Having lost its moisture during the ascent, the Chinook begins its downward journey as a dry wind. Now dry air is warmed by compression as it sinks (and is cooled if it rises) twice as rapidly as wet air is, i.e. at 1·6° F. per 300 feet. Fig. 28 explains how air at, say, 40° F. at the beginning of a 6,000 feet climb over the Rockies would finish its up and down journey with an increased temperature of 56° F.

So warm is the Chinook that in less than half an hour the temperature may rise by 40°; winter snows melt and are evaporated as if by magic, and summer pastures are turned into hay as they stand.

The föhn wind resembles the Chinook. In spring it melts the snows of the northern Alpine valleys of Switzerland. Occasionally two feet of snow have disappeared in a few hours. When the föhn blows the Swiss Alpine dwellers safeguard their wooden chalets by putting out their fires, for danger from sparks greatly increases in the extremely dry air.

TEMPERATURE AND THE HUMAN BODY

HUMAN comfort depends to a large extent upon the temperature of the air. In hot weather we perspire. This process is Nature's method of keeping us cool. The evaporation of moisture causes a fall in temperature, and when our perspiration evaporates we therefore feel cooler. The Negro in hot lands perspires very freely through his large pores and sweat-glands and thus automatically cools himself. By contrast, a white man who lives in these tropical regions remains uncomfortably hot, for his sweat-glands and pores are smaller and he cannot perspire so much as coloured peoples. His weakened body falls a prey to ailments, particularly stomach complaints.

The loss of this salty sweat naturally reduces the body's stocks of salt and water, but these are restored by eating salt and by drinking water. Salt is, in fact, such an absolute necessity in tropical lands that it is sometimes taxed to provide governments with an assured income.

The broad nose with wide nostrils, the thick lips with their large skin surface exposed to the air, and the short hair of the Negro all help to keep him cool by increasing both perspiration and the loss of bodily heat by radiation.

In hot regions a dark skin is Nature's way of protecting man from the ill effects of bright sunshine. To shield the tissues of the body from damage by strong sunlight, dark pigment forms beneath the skin. It is by this same process that we become sunburnt in a heat wave.

The Chinese, Japanese, and other Asians have yellow skins. The very cold, dry winters of the original homeland of these Mongolian peoples caused the body's sensitive blood vessels to be placed out of danger well beneath the skin, which itself was thick as a further protection against cold. The body consequently took on a sallow tinge. The almond shape of the Mongolian's eyes may have originated as a protection against the dazzling sunlight and strong, dust-laden winds of Central Asia.

The kinky hair of the Negro, the lank hair of Oriental peoples, the "pepper-corn" curls of the Hottentot, the fair hair of the North European—all of these and many other racial characteristics can be directly or indirectly traced to the effects of climate.

Since a coloured skin is largely the result of climate, coloured races should not be regarded as inferior to white peoples merely because they are black, brown, or yellow.

EXERCISES

1. Explain why the following weather sayings often prove to be correct:
   (*a*) Clear moon,
   Frost soon.
   (*b*) When the dew is on the grass,
   Rain will never come to pass.
   (*c*) When the grass is dry at morning light,
   Look for rain before the night.

2. In hot deserts anti-freeze mixtures are often needed at night in motor cars, and lost travellers have saved their lives by collecting and drinking dew. During the day-time maximum temperatures have broken world records. Why are desert nights so cold and the dews so heavy, and why are the days so hot?

3. "Posh" is derived from the initials P.O.S.H. ("Port Outward, Starboard Home") formerly printed on the most expensive return tickets for voyages from Britain to India. Show how latitude and aspect explain why the best cabins were reserved in this way.

CHAPTER VIII

How Temperature is Mapped

ISOTHERMS

IN order to show how mean temperatures change over a region we must not merely enter the figures in the correct places on an outline map. If we did so no clear-cut pattern would emerge from the maze of numbers. The figures are indeed plotted on the map in this way, but merely as a first step. Once the means have been placed in their right positions, all places which show the same figure are connected by a line called an *isotherm*

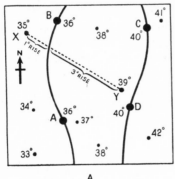 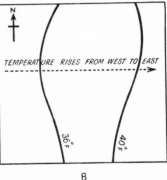

Fig. 29.—How Isotherms are drawn

(Greek *isos*=equal, *thermos* =heat). Fig. 29A illustrates how isotherms for 36° F. and 40° F. are drawn.

Notice how each isotherm is drawn not only right through all places which have the same temperature, e.g. A and B (both 36°) and C and D (both 40°), but also proportionally between places with temperatures above or below the figure required, e.g. the 36° isotherm goes between X (35°) and Y (39°) at a point which is three times as far from Y as from X.

The confusing jumble of figures can now be erased, leaving an isotherm pattern which clearly shows how temperature rises

[*Popper*

Snow in Africa

This snow scene in the Atlas Mountains of N.W. Africa
shows that temperatures are lower in highlands than in lowlands. When isotherm maps are drawn actual temperature
figures are adjusted to what they would be at sea-level. Are
there maps with actual temperatures in your atlas?

from west to east (Fig. 29B). We must not forget, however, that
even over a small area differences in altitude lead to great temperature changes. A lowland town is usually warmer than one
perhaps only a mile or two away but perched on a mountain-
side. In such cases isotherms based upon actual mean temperatures present a confusing maze of lines, and the important results
of latitude, distance from the sea, prevailing winds, and ocean
currents are hopelessly camouflaged by the single over-riding
effect of altitude. To overcome this difficulty the real means are
altered to what they would be if all places were at sea-level, and
isotherms are drawn through these sea-level figures. In this
process of "reduction to sea-level" we imagine that the place
is reduced, not the temperature, for this goes up as the place
"goes down" at an average rate of 1° F. per 300 feet.

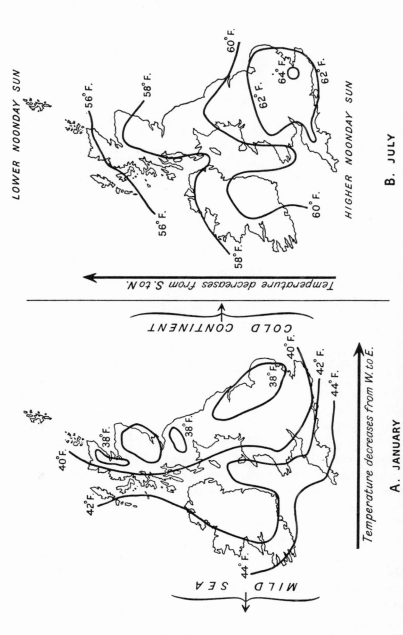

Fig. 30.—Isotherms over the British Isles

An isotherm may therefore be defined as a line which joins all places with the same mean sea-level temperature.

ISOTHERMS OVER BRITAIN

CONTROLLING sea-level temperatures over the British Isles are: (i) the Atlantic Ocean with its warm, on-shore North Atlantic Drift, and (ii) the noonday sun-angle. We will call them "sea" and "sun" respectively.

In January the general trend of isotherms is roughly north-

Fig. 31.—*The Temperature Quadrants of the British Isles*

south along the meridians, showing a west to east decrease in mean temperature from over 44° F. in South-west Ireland and Cornwall to below 38° F. in Eastern England and Scotland (Fig. 30A). The mild sea is now more important than the low sun, and the west, greatly influenced by the Atlantic Ocean, is milder than the east, which faces a cold continent. For instance, although both are on parallel 52° N., Killarney (43° F.) in South-west Ireland is milder than Cambridge (38° F.) in Eastern England. The south of England is commonly stated to be warmer than the north. This description is certainly true of

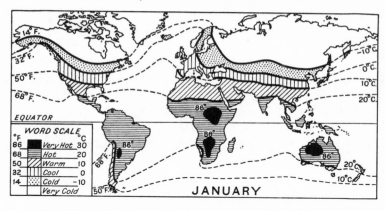

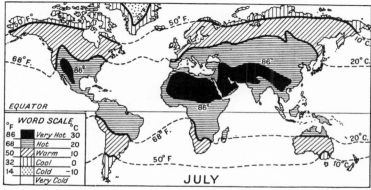

Fig. 32.—January and July Isotherms and Temperature Belts

the summer months, but is only partly so in winter, when Kent, for instance, being in the south-*east* is colder than Lancashire in the north-*west*.

In July British isotherms roughly follow the parallels, showing a decrease from over 62° F. in Southern England to below 56° F. in Northern Scotland (Fig. 30B). The high noontide sun, now playing a more important part than the sea, climbs higher in the "sunny south" than in the cooler north. Torquay (62° F.) in Devon is 7° warmer than Wick (55° F.) in North Scotland.

The "tug-of-war" of sea versus sun divides Britain into quadrants based upon differences in annual range of temperature (Fig. 31). Boundaries between the quadrants are pro-

82

vided by the isotherms of 40° F. (January) and 60° F. (July).

The greatest annual range occurs in the south-east quadrant, which is therefore the nearest British approach to a continental climate. Partly for this reason, East Anglia has become Britain's granary, for winter frosts break up the soil and kill insect pests, while summer warmth ensures a successful wheat harvest.

The smallest annual range (apart from South-west Ireland) is found in the north-western quadrant, where mild winters (at sea-level) are followed by cool summers.

ISOTHERM MAPS OF THE WORLD

THE patterns of isotherms showing sea-level temperatures over the world for January and July appear in Fig. 32. It is obvious how the position of each particular isotherm varies with the swing of the noonday sun, being much farther north in July than in January.

EXERCISES

1.

Quadrant	Place	January (*mean max.*)	July (*mean max.*)
N.W.	Douglas (54° 10′ N., 4° 28′ W.)	45° F.	64° F.
N.E.	Whitby (54° 29′ N., 0° 37′ W.)	42° F.	64° F.
S.W.	Bath (51° 23′ N., 2° 21′ W.)	45° F.	70° F.
S.E.	Tunbridge Wells (51° 8′ N., 0° 16′ E.)	42° F.	70° F.

Show how the mean maximum temperatures for the above places justify a division of Britain into temperature quadrants (Fig. 31). Explain carefully the reasons for the differences in temperature shown (*a*) in January and (*b*) in July.

2. Explain why the July isotherm of 60° F. in Fig. 30 sags equatorwards over the Irish Sea but bulges polewards over Ireland and England. Why does the January isotherm of 42° bend polewards over the Irish Sea?

3. Account for (*a*) the poleward sweep of the 32° F. (freezing-point) isotherm for January (Fig. 32) over the north-western margins of North America·and Eurasia, (*b*) its sag equatorwards over their interiors, and (*c*) its departure from their eastern margins at latitudes well south of those that mark its arrival over their western margins.

Why do no isotherms for 86° F. appear over the oceans in Fig. 32?

CHAPTER IX

Pressure Belts

WITHOUT rain all life would die. Whether or not it rains largely depends upon wind directions, which are in their turn controlled by changes in atmospheric pressure. In the study of climate and weather a knowledge of the great pressure belts and wind systems of the world is obviously of prime importance.

THE MEASUREMENT AND MAPPING OF PRESSURE

AIR, like everything else, exerts a pressure upon the land or sea upon which it rests. This pressure can be estimated by the height, in inches or millimetres, of a column of mercury supported by it in an inverted tube from which air has been exhausted (Fig. 33). The inch, however, was designed to measure length, and to measure pressure by it is really as absurd as to sell milk by the yard or to generate electricity by the pound. A special measurement, reserved solely for atmospheric pressure, has therefore been devised. This is called the *bar*, and pressure is recorded in millibars (1,000 millibars=1 bar).

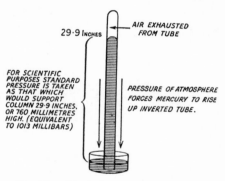

29·9 INCHES

AIR EXHAUSTED FROM TUBE

FOR SCIENTIFIC PURPOSES STANDARD PRESSURE IS TAKEN AS THAT WHICH WOULD SUPPORT COLUMN 29·9 INCHES, OR 760 MILLIMETRES HIGH. (EQUIVALENT TO 1013 MILLIBARS)

PRESSURE OF ATMOSPHERE FORCES MERCURY TO RISE UP INVERTED TUBE.

Fig. 33.—The Measurement of Pressure

Before a pressure map can be made the process of reduction to sea-level (as with temperature) must be carried out, 1 millibar increase in pressure being allowed for every 30-feet drop towards sea-level. Lines on a map which link all places of equal sea-level pressure, just as isotherms join places of equal sea-level temperature, are called *isobars* (Greek *isos*=equal, *baros*=weight).

WORLD DISTRIBUTION OF PRESSURE

THE atmospheric pressure of a region depends upon (1) its altitude, (2) its temperature, and (3) the rotation of the earth.

Altitude and Pressure.—Less of the atmosphere rests upon mountain-tops than upon lowlands. Consequently pressure decreases with increasing height above sea-level. Deafness, giddiness, mountain sickness, and a throbbing heart occasionally attack climbers in the rarified air of mountains.

Temperature and Pressure.—Warm air expands and becomes light but cold air is heavy. To be scientifically correct, however, we should call warm air "less dense" rather than "light" and cold air "dense" rather than "heavy". Warm air, being less dense, rises; if it did not tunnels instead of chimneys would be needed to get rid of smoke. The rising air flows outwards and pressure falls, for less air now rests upon the ground.

On the other hand, cold air, being dense, sinks. This descent is very noticeable in hilly regions on clear nights, when cold air from hill-tops drains down into valley-bottoms, filling them with misty pools of frosty air. Fruit growers, aware of this downhill "frost-drainage", plant orchards above valley floors, out of reach of the death-traps which in blossom-time can prove fatal. Cold air, therefore, by weighing down heavily upon the earth's surface, increases pressure.

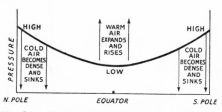

Fig. 34.—*Pressure Distribution as it would be if Temperature alone influenced Pressure*

If it were affected by temperature *alone*, pressure would therefore steadily increase from very low around the scorching Equator to very high at the icy Poles, as shown by the graph (Fig. 34).

Rotation and Pressure.—If a small marble is dropped on to a rotating gramophone turntable it will be thrown outwards from the slowly turning centre towards the more rapidly moving rim. In a somewhat similar way air, owing to the earth's rotation,

85

should be swept away from the slowly twisting polar regions and banked up towards rapidly rotating equatorial latitudes. Polar regions, thus robbed of air, should experience low pressure, while high pressure should prevail where air accumulates towards the Equator. In actual fact, conditions are far more complicated than this simple account suggests. Nevertheless, a graph showing pressure distribution as determined by rotation *alone* would be something like that given in Fig. 35.

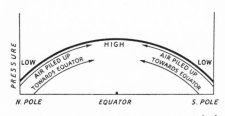

Fig. 35.—Pressure Distribution as it might be if the Earth's Rotation alone influenced Pressure

Obviously this graph is the reverse of that showing the effect of temperature. Both of these contradictory graphs cannot be everywhere correct, for the Equator and Poles cannot at the same time experience both high and low pressure. A struggle thus arises between temperature and rotation to control the distribution of pressure over the earth. Fig. 36 illustrates in diagram form this "tug-of-war". Notice how (*a*) *Temperature* gains control at the very hot Equator, where heat gives rise to

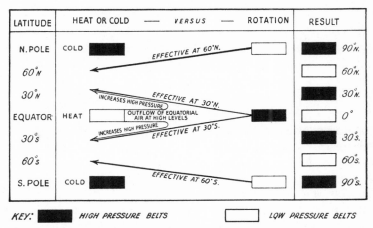

Fig. 36.—The "Tug-of-War" between Temperature and Rotation to control Pressure Distribution

low pressure, and also at the Poles, where intense cold causes pressure to be high; (*b*) *Rotation* brings its influence to bear elsewhere, approximately at 60° N. and S. and at 30° N. and S. Low pressure prevails at 60° N. and S., whereas at 30° N. and S. high pressure is brought about by the piling up of air in these latitudes. This high-pressure accumulation cannot be explained solely by rotation. It is largely due to the heaping up here of equatorial air which, warmed at the Equator, rises and flows outwards at high levels, to sink on cooling at about 30° N. and S.

Of course, to simplify this account of the distribution of pressure belts, it has been necessary to assume that the earth's surface consists entirely either of land or of water and that the noonday sun is at all seasons overhead at the Equator.

Fig. 37 shows how these belts of high and low pressure would appear on the globe, with an alternating rhythm of L.P.—H.P.—L.P.—H.P. at 0°—30°—60°—90° respectively. Note that to some of these pressure belts names are given. The low-pressure belt straddling the Equator forms the *Doldrums*, and the 30° N. and S. high-pressure systems are

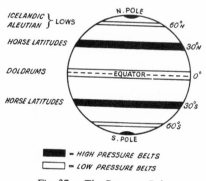

Fig. 37.—*The Pressure Belts*

called the *Horse Latitudes*. Named after islands, the low-pressure belt over the North Atlantic is known as the *Icelandic Low* and over the North Pacific as the *Aleutian Low*.

SEASONAL DISTURBANCES OF THE PRESSURE BELTS

WITH the exception of the polar high-pressure systems, these belts are not rigidly fixed throughout the year. Only in spring and autumn is the pressure pattern pivoted upon the Equator and spread out evenly on either side of it, as in Fig. 37. When spring fades into summer and autumn into winter, the design loses its symmetrical arrangement for two reasons:

1. *The Swing of the Sun.*—As the overhead sun swings from

Tropic to Tropic, the Doldrums low-pressure belt that is caused

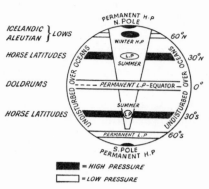

by its scorching heat naturally follows it. The Doldrums, however, lag behind their solar creator, and accompany the sun for not more than 10° of its 23½° migration on each side of the Equator. Since the Doldrums belt forms the central keystone of the structure of pressure systems, the belts on either side of it sway with it in its seasonal travels.

Fig. 38.—How Continents disturb the Pressure Belts

2. *Seasonal Changes of Temperature over Continental Interiors.*—Except for their seasonal swing with the sun, pressure belts are not disturbed over oceans, for the annual range of temperature there is comparatively small. Low pressure remains low and high remains high.

On the other hand, so close is the link between temperature and pressure that wherever continents breach the continuity of the oceans certain pressure belts are disrupted by the drastic continental change-over in temperature from summer to winter (Fig. 38).

For instance, wherever the Horse Latitudes cross land-masses their high pressure is in summer changed to very low by continental heat. The cooler winter brings conditions back to normal. Again, over the hearts of northern continents the bitterly cold winter transforms the 60° N. low-pressure into a huge high-pressure system. Here the normal low pressure not only returns as winter ends, but with summer heat sinks to a very low level indeed.

Naturally, pressure belts are interrupted most where continents are widest. Over the narrowing southern continents the swing from summer low to winter high pressure, although important, is not nearly so marked as in the northern hemisphere.

Not all pressure belts are disjointed over continents. The

Doldrums and the polar belts, for instance, continue round the world at all seasons since land and sea temperatures do not greatly differ in these latitudes. Again, no land reaches the 60° S. low-pressure system, which consequently encircles the world without interruption.

The pressure changes caused by the seasonal variations in temperature over continental interiors may be summarised as follows:

Pressure Belt	Normally	Whether upset	How	When	Why
Doldrums (0°)	Low	No	—	—	Always hot.
Horse latitudes (30° N. and S.)	High	Yes	High to low	Summer	Summer heat.
60° N.	Low	Yes	Low to high	Winter	Winter cold.
60° S.	Low	No	—	—	No continents.
Poles (90° N. and S.)	High	No	—	—	Always cold.

EXERCISES

1. The altimeter in an aeroplane registers an increase in height corresponding to the decrease in atmospheric pressure as recorded by a barometer. An aeroplane leaves an aerodrome where the sea-level pressure is 1,016 millibars and rises until the pressure is 910 millibars. Assuming that pressure falls by 1 millibar per 30 feet rise, how high is the aeroplane? If the pilot maintains this reading on his altimeter but sea-level pressure at his destination is 988 millibars, what is his real height at the end of his flight? A pilot is informed of the pressure changes which are likely to accompany weather changes during his flight, so that he can reset his altimeter. Suggest dangers which might be encountered if this information were not provided.

2. Why do pressurised cabins add to the comfort of passengers in high-flying aeroplanes? But for such devices, why would fountain pens leak and puffy food swell in the stomach?

3. Giving reasons, state whether pressure is high or low (a) at all seasons over the East Indies, Iceland, the Azores, and Tierra del Fuego; (b) in January over South Africa, Central Asia, Central Australia, and Central Canada; and (c) in July over Central Australia, North India, South Brazil, and the interior of North America.

CHAPTER X

The Winds of the World

LAND AND SEA BREEZES

SHOULD you be foolish enough to stick a pin into an inflated bicycle tube, you will soon learn that air blows from the high-pressure interior of the tube towards the lower pressure of the air outside of it. If the air flowed from low-pressure to high-pressure areas, then the more you punctured your tube the harder would it become!

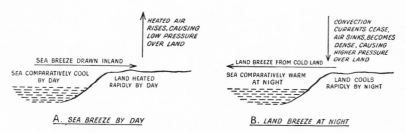

HEATED AIR RISES, CAUSING LOW PRESSURE OVER LAND

CONVECTION CURRENTS CEASE, AIR SINKS, BECOMES DENSE, CAUSING HIGHER PRESSURE OVER LAND

SEA BREEZE DRAWN INLAND

LAND BREEZE FROM COLD LAND

SEA COMPARATIVELY COOL BY DAY

LAND HEATED RAPIDLY BY DAY

SEA COMPARATIVELY WARM AT NIGHT

LAND COOLS RAPIDLY BY NIGHT

A. *SEA BREEZE BY DAY*

B. *LAND BREEZE AT NIGHT*

Fig. 39.—*Land and Sea Breezes*

That "the wind doth blow from high to low" and not "where it listeth" can also be proved when, on a hot summer afternoon at the seaside, you are fanned by a cool sea breeze. The land, absorbing heat more rapidly than the sea, warms the air above it, so that a local low-pressure system is set up, towards which breezes blow from the comparatively high-pressure region over the water (Fig. 39A).

After sunset these conditions are reversed. Air over the land, now rapidly cooling, is colder than that over the sea, which parts reluctantly with the heat that it slowly gained by day. High pressure develops inland and the sea breeze of the day gives way to a nocturnal land breeze (Fig. 39B).

Wind Direction and Ferrel's Law

Winds do not follow the shortest path from a region of high to one of low pressure, for Ferrel's Law (page 36) applies to winds as well as to all other moving bodies. The isobars in Fig. 40 show high-pressure and low-pressure systems in each hemisphere. Notice how the departure of the full lines (actual wind directions) from the dotted lines (direct course of winds from high to low pressure) agrees in every case with Ferrel's Law of

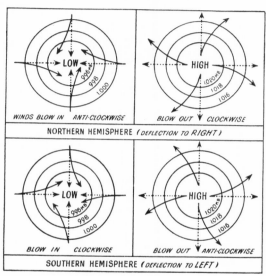

Fig. 40.—*Ferrel's Law applied to Winds*

deviation to the right in the northern and to the left in the southern hemisphere.

Trades, Variables, and Polar Winds

From the high-pressure "reservoirs" of the Horse Latitudes at 30° N. and S. winds stream away in opposite directions to fill up the low-pressure "troughs" of (i) the equatorial Doldrums and (ii) 60° N. and S.

(i) The air that is forced from the Horse Latitudes towards the Doldrums forms the *Trade Winds*. These winds maintain a

[*Popper*

The Effect of Trade Winds upon Vegetation

In some areas the trade winds are very steady. Their effect upon the shape of trees in parts of the West Indies is shown in this picture of Willemstad, in Curaçao. Which trade winds blow here?

steady track; in fact, the word "trade" comes from the Anglo-Saxon verb *tredan* = to tread, i.e. to follow a regular path.

(ii) Some of the air from the Horse Latitudes moves away in the opposite direction to the trades, towards the 60° N. and S. low-pressure belts. Unlike the trades, these winds are not very constant. They are particularly unsteady in the northern hemisphere, where seasonal pressure changes over the wide continents play havoc with their strength and direction. Blowing mainly from westerly directions, they are commonly called *Wester-lies*, but their fickleness has also earned for them the name *Variables*, and it is by this name that we shall know them in later pages.

(iii) A third set of winds blow from the polar high-pressure "reservoirs" to-

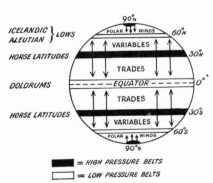

= HIGH PRESSURE BELTS

= LOW PRESSURE BELTS

Fig. 41.—Planetary Winds as they would blow over a Stationary World

wards the 60° N. and S. low-pressure "troughs". These are the *Polar Winds*.

In Fig. 41 these three groups of *planetary* winds are shown as they would blow if the earth were not rotating, i.e. directly from high- to low-pressure belts. Just as rivers flow downhill from a watershed, so these winds blow down the *pressure gradient* from high to low levels.

The earth's rotation upsets this simple wind pattern. Applying Ferrel's Law and remembering that winds are named according to the direction *from* which they blow, we find (Fig. 42) that in the northern and southern hemispheres respectively: (i) the

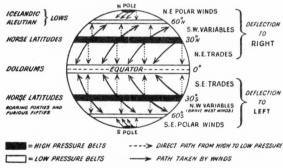

Fig. 42.—Trades, Variables, and Polar Winds

trades become north-east and south-east, (ii) the variables are south-west and north-west, and (iii) the polar winds, like the trades, blow from the north-east and south-east. The north-west variables (southern hemisphere) sweep round the world practically uninterrupted by continents, reaching such a force that they are sometimes called the *Brave West Winds*. The latitudes through which they blow are described as the *Roaring Forties*, *Furious Fifties*, and the *Shrieking Sixties*.

These wind-systems may be easily remembered as follows:

1. The final letter E of "tradE" suggests east. Place it before the word "trades"—"E. trades". In front of E now put the initial letter of the hemisphere concerned, i.e. N for the Northern and S for the Southern hemisphere. The correct directions of the trade winds will then appear—N.E. and S.E.

2. In each hemisphere reverse the compass directions of the trades to find the variables, i.e. S.W. (opposite to N.E. trades) and N.W. (opposite to S.E. trades).

3. In each hemisphere give the polar winds the same direction as the trades, i.e. N.E. and S.E.

THE ATMOSPHERIC "BOILER".

So far we have considered only surface winds. We have yet to see how the high-pressure "reservoirs", from which air is constantly draining away, replenish their stocks.

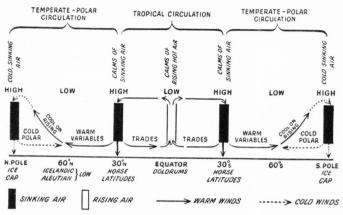

Fig. 43.—*The Atmospheric "Boiler"*

Inside a boiler hot water continually moves round and round in a well-marked circulation. Air circulates in a somewhat similar way, with the high sun within the Tropics providing a "furnace" for the atmospheric "boiler". There are, in fact, two distinct circulations in each hemisphere: (1) a tropical circulation operating between the Horse Latitudes and the Equator, and (2) a temperate-polar circulation between the Horse Latitudes and the Poles.

1. *The Tropical Circulation.*—In the Doldrums the air, heated by the solar "furnace", becomes less dense and rises (Fig. 43). Sooner or later it must, at high levels, spread outwards on either side towards the Poles. The Doldrums, some 600 miles

94

wide, form calms, for whenever air rises or sinks it cannot be felt as a wind. Most of this outflowing air up aloft, instead of continuing all the way to the Poles, is forced by the earth's rotation to sink to the surface at the Horse Latitudes. The sinking air of the Horse Latitudes forms calms, just as the rising air of the Doldrums does.

On reaching the ground, some of the air flows back towards the Doldrums as the trade winds. There is, however, no "head-on collision" at the Equator between the two converging trade wind systems. Warmed by the solar furnace, they merge

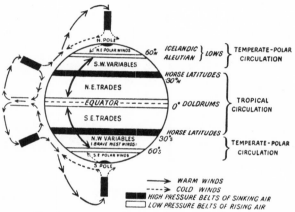

Fig. 44.—Summary of Atmospheric Circulations

into the rising air column of the Doldrums and so complete the tropical section of the atmospheric "boiler".

2. *The Temperate-Polar Circulation.*—Not all of the air that descends at the Horse Latitudes returns to the Doldrums. Some of it, forming the variables, is forced polewards towards the 60° low-pressure belts (Fig. 43).

Around 60° this warm, light, tropical air meets the cold, dense, polar winds blowing outwards from the polar high-pressure ice-caps. Riding up over these dense polar blasts, the lighter variables continue their poleward journey at high levels until, losing their heat, they sink at the Poles as cold, heavy air. Reaching the surface, this air streams back towards the 60°

low-pressure "troughs" as polar winds, thus completing the temperate-polar section of the atmospheric "boiler".

The tropical and temperate-polar circulations are obviously linked to each other at the Horse Latitudes (Fig. 43). Here the sinking air forks, some to take part in the one circulation and the rest in the other.

In Fig. 44 the diagram of the atmospheric circulation is added to the surface-wind pattern of Fig. 42.

SEASONAL DISLOCATIONS OF THE WIND SYSTEMS

THIS symmetrical pattern of wind systems pivoted on the Equator could be maintained throughout the year only in a world wholly covered by either land or water, with a noon sun always overhead at the Equator.

Since winds obey pressure, variations in pressure are accompanied by changes in wind direction. The regularity of the world's pressure pattern is upset by (i) the migration of the overhead sun, and (ii) the great range between summer and winter temperatures over continental hearts (Chapter IX). Consequently the wind pattern is broken up for the same two reasons. Seasonal dislocations of wind systems caused by the migrating sun are referred to in Chapter XII. At the moment we will consider those resulting from the great annual range of temperature over continental interiors.

Over land-masses a breakdown in wind systems follows the breakdown in pressure belts, i.e. in winter over the wide northern continents in the 60° N. low-pressure belt and in summer over the high-pressure belts of the Horse Latitudes (page 88). So important in the study of climate are these great flaws in the regularity of the wind pattern that they merit special examination.

The Continental "Lungs" of North America and Eurasia.— Over the heart of North America and Eurasia there develops in winter a huge "flywheel" of outblowing winds. From the high-pressure hub of this "wheel", caused by intense cold, winds are flung out and deflected to their right into a great clockwise swirl (Fig. 45A). In mid-winter this high-pressure

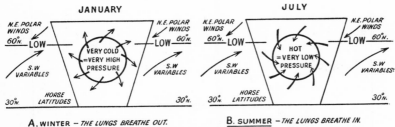

Fig. 45.—The "Continental Lungs" of Northern Land-masses

system sometimes reaches out from Eurasia as far as Britain, bringing biting east winds and snowstorms which persist until the mild on-shore variables from the Atlantic resume their sway and force the invading continental air-masses to retreat.

In summer the low-pressure "trough" of 60° N. is restored. It is, in fact, greatly enlarged and deepened by continental heat. From all sides winds are drawn in towards it and are deflected to their right into an anti-clockwise swirl (Fig. 45B).

These continental heartlands may be suitably compared to huge lungs, gulping in streams of air in summer and forcing them out in winter.

Tropical Monsoons.—Monsoon winds (Arabic *mausim*=season) are caused in much the same way as land and sea breezes. In both cases air is attracted inland by low pressure following intense heat, and is later on thrust out to sea during cooler weather. For their development, however, monsoon winds need much more space and time than do coastal breezes, for thousands. of square miles of continents and oceans are involved, and instead of merely a day and a night a whole year is required,. with a very hot summer followed by a cooler winter.

The operation of the monsoon system is best seen in the great seasonal interchange of air-masses between South-east Asia (especially India) on the one hand and the northern coast-lands of Australia on the other. India's frequent famines and Northern Australia's loss of nearly 1,000,000 cattle in the summer drought of 1951–52 show what disasters often follow a failure in these rain-bearing winds.

[*E.N.A.*

Winter Drought in India

This *nullah*, or dried-up watercourse, shows that drought often occurs in India in the dry season, when the N.E. monsoon winds are off-shore.

THE INDIAN AND AUSTRALIAN MONSOONS

(i) *Around January* (Fig. 46A).—In winter over India and South-east Asia the normal Horse Latitudes high-pressure belt reigns, with outblowing off-shore north-east trades. Most of India is now dry. At the Equator the low-pressure Doldrums persist, but over Central Australia the high pressure of the southern Horse Latitudes is now transformed by summer heat into very low pressure. Pressure thus falls steadily all the way from India (high) past the Doldrums (low) to Australia (very low). From South-east Asia the north-east trades, now called north-east monsoons, are forced across the Equator into the southern hemisphere. On arriving here they are deflected to the left. Consequently they reach Northern Australia as on-shore north-west monsoons. Northern Australia receives the benefit of their long sea journey in the form of torrential summer rains which, however, rapidly decrease inland.

(ii) *Around July* (Fig. 46B). Conditions are now reversed. Over North India summer heat converts the Horse Latitudes high-pressure zone into an area of very low pressure. In the Dol-

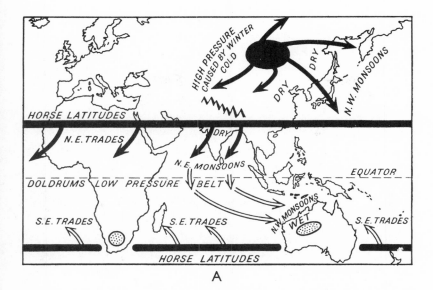

A

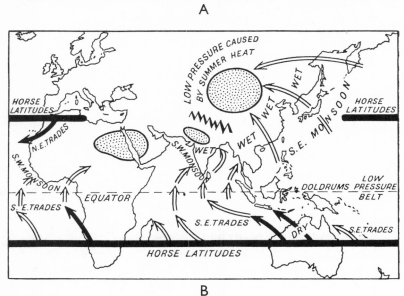

B

Areas of High Pressure — Dry, off shore winds

Areas of Low Pressure caused by summer heat — Moist winds from sea

Mountain barrier of Himalayas

Fig. 46.—Monsoon Winds

99

[*O. H. Borradaile*

Summer Rain in India

In summer the S.W. monsoon winds blow on-shore over India, bringing rain. Note the rain-clouds and evidence of a recent deluge.

drums pressure is low, but over Central Australia (winter) the southern Horse Latitudes high-pressure belt resumes control.

Like water running downhill, winds blow down this steady pressure gradient from the high of the southern Horse Latitudes past the low of the Doldrums to India's very low pressure. They set out on their journey as the normal south-east trades, off-shore from Northern Australia, which now has its winter drought. Forced across the Equator towards the very low pressure over North India, they are deflected to the right once they enter the northern hemisphere. Consequently they finish their round-about sea voyage as thoroughly moist south-west monsoons. Rising against the mountain walls of India and South-east Asia, they deluge the land with life-giving summer rains.

"As the monsoon draws near, the days become more over-cast and hot and banks of cloud rise over the ocean to the west. . . . At last the sudden lightnings flash among the hills,

[*Fox*

The River Nile in Flood in Egypt

Heavy monsoon rains in summer in Abyssinia are the chief cause of the Nile floods, which are all-important to the Egyptian peasant.

and sheet through the clouds that overhang the sea, and with a crash of thunder the monsoon bursts over the thirsty land, not in showers or partial torrents, but in a wide deluge that in the course of a few hours overtops the river banks and spreads in inundations over every level plain. . . . The rain descends in almost continuous streams, so close and so dense that the ground is covered with one uniform sheet of water. For hours together the noise of the torrent occasions an uproar that drowns the ordinary voice, and renders sleep impossible."—Sir J. E. Tennent.

OTHER MONSOON LANDS

CHINA and Japan form a temperate counterpart to the tropical monsoon lands of South-east Asia (Fig. 46A and B).

Here there is in winter a similar outflow of dry winds from the continental high-pressure "lungs", in this case centred over the heart of Asia. Japan, however, receives some precipitation in winter, for these outblowing winds must cross water before they reach it and so they become on-shore. Winters in these East Asian margins are much colder than in tropical monsoon

101

lands, for the outblowing winds come from the icy continental heart, whereas the Himalayas prevent them from reaching India.

In summer there is the same reversal of winds as in tropical monsoon regions, and torrential rains are brought ashore by moist, on-shore winds, drawn in from the Pacific towards the low-pressure "lungs" over the heated continent. Over these temperate monsoon lands the winter outflow of air is from the north-west, while the summer inflow is from the south-east.

The Guinea Coast of West Africa is also affected by monsoon winds. Around July south-east trades flow out from the South African (winter) Horse Latitudes high-pressure system and are pushed across the Equator towards the intensely hot Sahara Desert, arriving on the Guinea Coast as south-west monsoons (Fig. 46B). In fact, rain-bearing south-westerly winds here blow on-shore for much of the year. The Guinea Coast, moreover, is unlike India and Northern Australia in that it is near enough to the equatorial all-seasonal rainfall belt to escape the long winter drought. Another African region to receive monsoon winds is Abyssinia, whose summer rains, by causing floods to sweep northwards along the Nile, throw a life-line across the Sahara Desert to rainless Egypt.

EXERCISES

1. Find the extent in latitude of the following regions and, stating whether they are on-shore or off-shore, give the full name of the prevailing winds which blow over each: Kalahari Desert, British Columbia, West Indies, Patagonia, British Isles, Sahara Desert, South-east Brazil, South Chile, Desert of West Australia, Arizona Desert, South-east Queensland, South Island of New Zealand, Atacama Desert, Natal, and Tasmania.

2. Find out (*a*) how the monsoon winds influenced the first voyage of Vasco da Gama, and (*b*) the time of year at which many Arab dhows can be seen in harbour at the East African ports of Mombasa, Dar-es-Salaam, and Zanzibar.

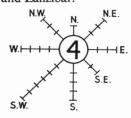

3. The diagram shows a monthly *wind-rose*. The "spokes" are drawn to scale according to the number of days on which the wind blows from the directions indicated. The number of calms appears in the circle. Record the wind direction at 9 a.m. on every day of the month and draw monthly wind-roses for as many months as possible. What do they tell you about the prevailing winds in your district?

CHAPTER XI

Why it Rains

RAINFALL is collected in a rain-gauge and is measured in milli-metres or inches, but when rainfall maps are made the figures, unlike those for temperature and pressure, cannot be corrected to sea-level conditions. In fact, far from wishing to ignore the effect of altitude upon rainfall, we want to find out how moun-tains govern its distribution. On a map lines that join places of equal rainfall are called *isohyets* (Greek *isos*=equal, *hyetos*= rain).

WATER-VAPOUR AND CONDENSATION

Water-vapour.—There is no such thing as completely dry air, for even in deserts moisture is always present in the gaseous form of water-vapour. Under certain conditions this invisible vapour takes shape as the tiny water droplets of a cloud or as rain, snow, hail, fog, or dew, all of which are covered by the word "precipitation". Whenever this transformation takes place the water-vapour is said to be condensed.

There is a limit to the quantity of water-vapour that air can absorb. This limit, however, varies with the ever-changing tem-perature of the air. Warm air can hold more water-vapour than cold air, and as its temperature rises air absorbs moisture at an ever-increasing rate. The table below gives the increase in moisture-holding capacity of 10 cubic yards of air (weight 22 lb.) as its temperature rises from 50° F. to 90° F.

Temperature	Water-vapour Limit		Increase per 10° Rise
50° F.	2·5 oz.		
60° F.	3·6 ,,	per 22 lb.	1·1
70° F.	5·0 ,,	of air (10	1·4
80° F.	6·8 ,,	cubic yds.)	1·8
90° F.	9·3 ,,		2·5

Dew-point and Condensation.—When you hold a carton of ice-cream you will find that it is soon covered with a film of water. What has happened?

Air which is holding water-vapour up to the limit for its temperature is said to be saturated. This particular temperature is called dew-point, for if the air is cooled condensation must begin and some of the water-vapour will be converted into waterdrops.

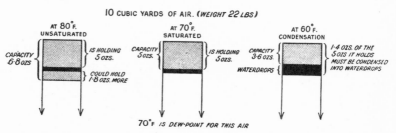

Fig. 47.—*The Effect of Cooling upon the Moisture-bearing Capacity of Air*

For example, if 10 cubic yards of air is holding 5 ounces of water-vapour at 80° F. it is not saturated, for its limit at this temperature is 6·8 ounces. If it is cooled to 70° F. it becomes saturated, for at this reduced temperature it can hold 5 ounces and no more. Should it now be further chilled to 60° F., its limit dwindles to 3·6 ounces and 1·4 ounces of its original store of 5 ounces must be condensed into visible waterdrops (Fig. 47).

For this particular sample of air 70° F. is the dew-point, for it marks the stage at which condensation begins. To "throw overboard" a mere 1·4 ounces does not seem to be a great sacrifice on the part of the 10 cubic yards of air, but the total amount of rainfall is often colossal. For instance, a fall of one inch upon a full-size football pitch weighs about 270 tons.

For condensation to take place a tiny speck of solid matter is normally required, around which the waterdrop can form. In the atmosphere countless millions of dust particles serve this purpose, and it is no exaggeration to say that without dust we should die, for within each life-giving raindrop lies a dust core.

We can now understand why the ice-cream carton feels damp. Air in contact with the cold surface is chilled until its tempera-

[*Popper*

The Condensation of Water-vapour

Here warm air is being chilled by a very cold window-pane until dew-point is reached. In this case the water-vapour is condensed as frost, since the dew-point is below 32°.

ture falls below dew-point, and its water-vapour is condensed upon the carton as tiny waterdrops.

In the same way dew "falls" during clear nights, when air is chilled by a rapidly cooling earth (page 72). In cloudless deserts many plants and insects depend upon the remarkably heavy night dews rather than upon rain.

How Air is Cooled

RAINFALL, then, follows a fall in temperature, for whenever and wherever it rains the fundamental cause is the cooling of air containing water-vapour. We now have to find out how this cooling of air takes place.

Cooling by Contact with a Cold Surface.—Air over a cold surface, e.g. a cold ocean current, may be chilled until its temperature falls to below dew-point. Condensation taking place under these circumstances creates mists and fogs rather than rain.

The Evaporation of Water
[L.N.A.

A cloud is here seen descending from Table Mountain (Cape Town). Warmed by compression, the air can absorb more moisture and the cloud is dispersing as its water droplets are evaporated.

For instance, the western coasts of hot deserts in trade-wind latitudes are washed by cold currents. Here sea breezes, drawn ashore by summer heat, bring no rain to relieve the parched land. Out at sea they are chilled sufficiently by the cold water to cause mists and clouds, but, once ashore, they are heated by the hot land so that condensation ceases and drought persists. These *cold-water coasts* are sometimes fog-bound for weeks at a time.

Again, whenever a warm current approaches a cold one warm air over the former is cooled as it mingles with cold air over the latter. The dense coastal fogs off Newfoundland, where the warm North Atlantic Drift draws near to the cold Labrador Current, are formed in this way.

Cooling by Ascent.—Whenever air rises it expands on being surrounded by air at a lower pressure than itself. This expansion causes it to cool. Ascending air is therefore accompanied by cloud and rain. Conversely, descending air currents are warmed by compression as they approach the ground, where atmospheric pressure is higher than it is up aloft. The increasing warmth of these sinking currents enables them to absorb more water-vapour, and clouds rapidly disappear as their water droplets revert to invisible vapour. This process is well illustrated by the warm Chinook and föhn winds (pages 75 and 76).

Rainfall is given three names according to the three different ways in which air is forced to rise:

(*a*) *Convectional Rain.*—Hot air ascends as it becomes less dense. The rising air-mass is called a convection current. Such currents reach their climax in huge "tower" or "heap" clouds and in torrential thunderstorms. Occasionally in hot deserts these cloudbursts transform dried-up watercourses into torrents that overwhelm and drown unwary travellers. The heat-wave type of thunderstorm in Britain is convectional.

(*b*) *Orographic, or Relief Rain.*—Orographic rain (Greek *oros* = mountain) pours steadily from clouds caused by the cooling of air as it blows up mountain-sides. Holiday-makers know only too well how wet the Lake District, the Welsh Mountains, and the Scottish Highlands can be. Naturally, the area to the leeward of such wet windward slopes is comparatively dry. This drier region is said to be in the *rain-shadow* of the mountains. Thus, while the Andes of South Chile are drenched with relief rain from on-shore westerly variables, Patagonia, to their leeward side, is a semi-desert. Yorkshire, in the rain-shadow of the Pennines, is drier than windward Lancashire.

(*c*) *Rain from Depressions.*—The variables literally "run up against" the polar winds around 60° N. and S., for the warm, light, moist winds from the Horse Latitudes ride up over the cold, dense, dry air from the Poles (Fig. 53). Both layer and heap clouds occur, according to whether the warm air rises gently or is violently forced upwards (Fig. 55). These encounters between warm and cold winds are known as *depressions*.

RAINFALL IN BRITAIN

BRITISH weather is often the subject of good-humoured comment among its victims. Shakespeare notes that "The rain it raineth every day"; Lord Byron wittily stated that "The English winter ends on June 30th, and begins again on July 1st"; and

107

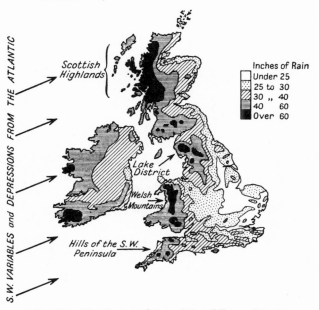

Fig. 48.—*Distribution of Annual Rainfall over Britain*

Note 1. The General decrease from West to East.
　　2. The Mountainous areas of the West receive most rain.

Coleridge once observed, "The English summer has set in again with its usual severity".

It is true that rain falls throughout the year in the British Isles, either from depressions or as relief rains over highlands. Even low hills such as the chalk Downs of South-east England receive relief rain, while no part of the country, highland or lowland, escapes the attentions of depressions.

Western regions of Britain are the first to welcome depressions, comparatively young and vigorous, from their Atlantic birthplace. Consequently the west of the country is wetter than the east. Moreover, since the highest mountains are in the west, heavy relief rain from the on-shore south-west variables increases the western total (Fig. 48). For instance, in the Lake District 130 inches of rain fall annually at Seathwaite. By contrast, on the same parallel (54° N.) but on the east coast lies Whitby with only 25 inches.

[*Camera Press*

A Convectional Thunderstorm

When convection currents over a heated land surface are very strong they give rise to thunderstorms. These are frequent in tropical lands and in summer in continental interiors. This picture was taken in the interior of the United States.

Convectional rain falls in Britain only after a period of hot weather. Sometimes many days of heat elapse before the convection currents are powerful enough to furnish a thunderstorm, whereas in the Doldrums this stage would be reached practically every afternoon.

EXERCISES

1. Air is holding 3 ounces of water-vapour per 10 cubic yards at X, 4 ounces at Y, and 2 ounces at Z. The temperature at all three places is at first 70° F., but falls to 50° F. Using the table on page 103, state (*a*) how many times heavier the rainfall is at Y than at X, and (*b*) how much water-vapour per 10 cubic yards must still be absorbed at Z before rain can fall.

2. The odds against a thunderstorm on a summer day are 9 to 1 at Cambridge, but 178 to 1 in the Shetland Isles. Why are the odds shorter at Cambridge?

3. If you had to choose a British seaside resort for a summer holiday as warm, sunny, and dry as possible, suggest a suitable choice, giving your reasons.

CHAPTER XII

A World Pattern of Cloud and Rain

IN Chapter VI various climates were described in so far as they differ from one another in temperature. Rainfall, however, plays a no less important part than temperature in forming the world's different climates.

When discussing climate, we should always remember that man's artificial political divisions rarely tally with Nature's climatic regions. Different kinds of climate and natural vegetation merge gradually into one another, and in most cases it is impossible to tell where one type ends and the next begins.

THE PATTERN OF CLEAR AND CLOUDY SKIES

A MAN in the moon, looking at our planet through a powerful telescope, would see belts of cloud alternating in a definite pattern with clear skies (Fig. 49):

(i) The convection currents of the Doldrums would veil equatorial lands with thick blankets of cloud and rain.

(ii) Beyond the Doldrums the earth's surface would appear, and skies would become particularly clear in the Horse Latitudes, where descending, warming air makes condensation impossible. For this reason many of the world's deserts straddle the Horse Latitudes.

(iii) Polewards from the Horse Latitudes the amount of cloud would gradually increase, reaching a climax around the 60° N. and S. low-pressure belts, where the clouds of depressions would once more blot out the surface. In the northern hemisphere, however, the dry continental interiors, far removed from the rain-swept coastal margins, would be seen through holes torn in this cloud blanket of temperate latitudes.

(iv) Little reliable information about polar regions is available. Theoretically, the lunar observer should be able to see the polar

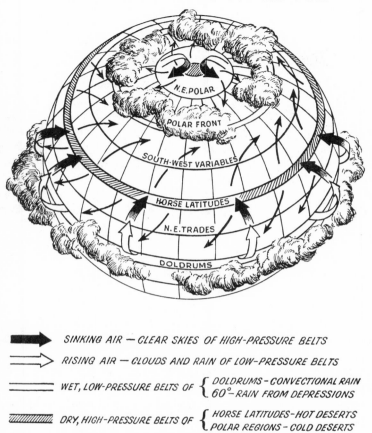

SINKING AIR — CLEAR SKIES OF HIGH-PRESSURE BELTS

RISING AIR — CLOUDS AND RAIN OF LOW-PRESSURE BELTS

WET, LOW-PRESSURE BELTS OF { DOLDRUMS – CONVECTIONAL RAIN
                              { 60°– RAIN FROM DEPRESSIONS

DRY, HIGH-PRESSURE BELTS OF { HORSE LATITUDES–HOT DESERTS
                              { POLAR REGIONS – COLD DESERTS

Fig. 49.—A Pattern of Clear and Cloudy Skies

ice-caps, for cold air cannot contain much water-vapour and at the Poles themselves the air is sinking, so that the amount of condensation should be small. In actual fact, however, the earth would probably often be veiled from view by frequent mists or by blizzards caused by strong winds whipping snow from the ground, such as those which overtook Captain Scott and his companions in their ill-fated retreat from the South Pole.

Many flaws interrupt this rhythm of alternating clear and cloudy belts. Nevertheless, the pattern of cloudy-clear-cloudy-clear skies corresponding to rising-sinking-rising-sinking air currents does give a general picture of what might be expected from the workings of the atmospheric "boiler".

The Pattern of Rainfall

The distribution over North and South America of the three kinds of rainfall—convectional, relief, and rain from depressions —is shown in Fig. 50. Over the remaining continents the pattern is similar, although here and there it is upset at certain seasons, especially over monsoon Asia, whose "lungs" in summer breathe in rain-bearing winds from the oceans but in winter breathe out dry air (see pages 96 to 102).

Convectional rains occur at all seasons throughout the Doldrums and in summer over heated continental interiors (Fig. 50A). Relief rain, as shown in Fig. 50B, falls wherever prevailing winds are on-shore, as where variables blow towards temperate maritime and trades towards tropical maritime regions. Depressions bring rain to both eastern and western margins in temperate latitudes but are unknown in tropical lands (Fig. 50C).

If a climatic region lies fairly and squarely well within a particular wind or pressure belt, its seasonal distribution of rainfall is in no way affected by the north-south sway of the wind-systems with the migrating overhead sun. For instance, temperate maritime regions like North-west Europe, including the British Isles, receive rain at all seasons from on-shore variables and depressions, no matter how much these winds follow the north-south swing of the sun. Similarly, equatorial lands never escape the influence of the Doldrums and are therefore deluged from convectional thunderstorms throughout the year. Hot deserts, too, at all times lie where trade winds are off-shore and therefore dry.

On the other hand, there are certain regions where the seasonal distribution of rainfall is closely bound up with the swing of the wind and pressure belts. Such *transitional* regions lie approximately in between two wind-systems. A transition means a passage, or change-over, from one set of conditions to another set of conditions, and a transitional region thus marks a half-way stage between two regions of widely different climates and usually of different types of natural vegetation.

AREAS RECEIVING RAIN FROM:

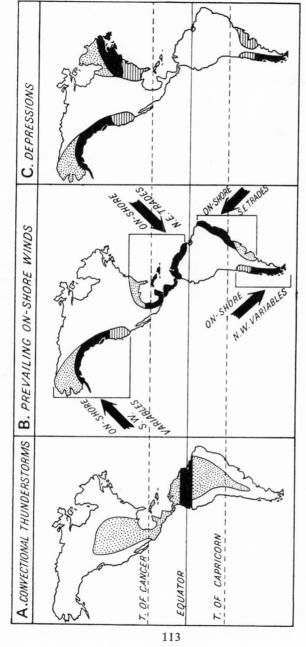

Fig. 50.—*Seasonal Distribution of Types of Rainfall over the Americas*

113

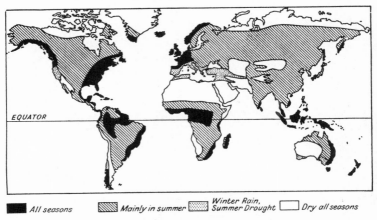

All seasons   Mainly in summer   Winter Rain, Summer Drought   Dry all seasons

Fig. 51.—The Seasonal Distribution of Rainfall

As the fortunes of war ebb and flow, the "no-man's-land" between two opposing armies is invaded by each army in turn. Similarly, a transitional region forms a "no-man's-land" situated between two wind-systems. It is covered first by one system and then by another as the wind and pressure belts shuttle to and fro in the wake of the migrating sun. For instance, over western warm temperate margins the *Mediterranean* climate reigns. To these lands on-shore variables and depressions bring rain in winter. As spring approaches, however, the variables and depressions retreat polewards, and by summer their place has been taken by the dry calms of the Horse Latitudes and by the equally dry off-shore trade winds which blow out from them. The Mediterranean climate is therefore one of winter rain and summer drought (see page 184).

It should be understood that although wind, pressure, temperature, and rainfall belts accompany the sun in its swing, they cannot keep pace with it, but lag behind it both in time and in space, especially over the oceans and over those margins of continents that feel oceanic influences. There is a lag in time, for whereas the overhead sun arrives at the northern and southern termini of its journey by June 21st and December 22nd respectively, these climatic belts do not reach the limits of their swing until a month or more later, namely by July-August and

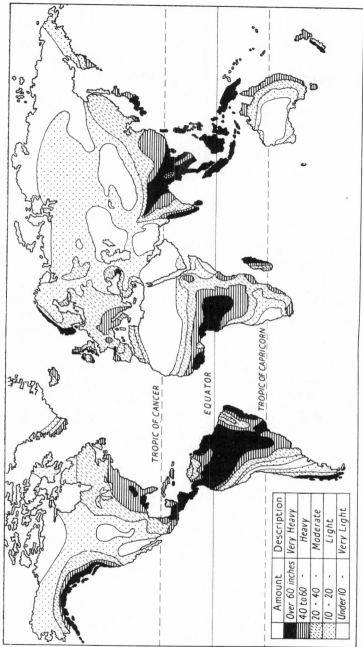

Amount	Description
Over 60 inches	Very Heavy
40 to 60 "	Heavy
20 - 40 "	Moderate
10 - 20 "	Light
Under 10 "	Very Light

TROPIC OF CANCER

EQUATOR

TROPIC OF CAPRICORN

Fig. 52.—Mean Annual Amount of Precipitation

January-February. Similarly, there is a lag in space, for the overhead sun moves through 47° of latitude, from 23½° N. to 23½° S., whereas the climatic belts, except for temperature zones over the interiors of the wide northern continents, do not stray for more than 5° to 10° northwards and southwards from their mean positions, i.e. through approximately 10° to 20° of latitude at the most. Over the open oceans the full swing is rarely through more than 10° and often through much less.

The annual amount of precipitation over the world is shown in Fig. 52, and the seasonal distribution in Fig. 51.

EXERCISES

1. North-western Europe receives on-shore variables and depressions throughout the year, whereas Southern Europe is visited by them in winter only. Explain why. What wind- and pressure-system prevails over Southern Europe in summer? Describe the seasonal distribution of precipitation which results from these seasonal changes in wind and pressure belts over Southern Europe. What name is given to this climate, and what other parts of the world in similar latitudes experience the same kind of climate?

2. By comparing Fig. 52 with Fig. 72 (page 217) find as many resemblances as you can between the pattern of annual rainfall and that of density of population over the world.

3. Explain (a) why in the Amazon Basin, the Congo Basin, and the East Indies (except Java) the density of population is small in spite of an abundant and well-distributed rainfall, and (b) why a long, narrow, densely populated strip crosses the desert in Egypt.

CHAPTER XIII

War and Peace in the Air

As we have already learned, much of the rain that falls in temperate lands comes from depressions.

A depression may be likened to an aerial battle, waged along a battle-line hundreds of miles long between two great air-masses weighing millions of tons. In fact, military terms such as *front* and *sector* are used to describe certain parts of a depression. The aerial "war" consists of a series of rainy "battles" formed by the depressions. Between each battle occurs a short lull of fine weather known as a ridge of high pressure. At length the war of the depressions, which usually lasts for several weeks, is followed by the peace of the anticyclone, a long period of dry, settled weather.

THE BATTLE OF THE DEPRESSION

The Two Air Forces.—Where do these aerial armies come from? Forming one army are the westerly variables, flowing polewards from the high-pressure calms of the Horse Latitudes. We may label these warm winds the *tropical* air force. Streaming equatorwards from high latitudes to block the advance of this tropical air are the icy blasts of the polar winds, which we will call the *polar* air force (Fig. 53). Polar air does not necessarily come from the polar ice-caps. Any air-mass originating from high latitudes is called polar—in fact, those from the ice-caps themselves are more correctly styled *arctic* air.

An attack by a tropical air-stream upon polar air is the "trigger" that starts the war. From small beginnings the struggle grows into a conflict whose course modern man can with great accuracy forecast, though he is quite powerless to control it. The following chart shows how these two great air opponents differ in every possible way:

117

	Tropical Air	*Polar Air*
Origin	Horse Latitudes	High latitudes, including polar ice-caps and (in winter) continental interiors.
Temperature	Warm air	Cold or cool air.
Density	Less dense (because warm)	Dense (because cold).
Humidity	High (i.e. moist)	Low (i.e. comparatively dry), except when it crosses much sea.
Mode of "Attack"	Steady, invasion of polar air	Violent squally attack on tropical air.
Direction	Moves polewards	Moves equatorwards.

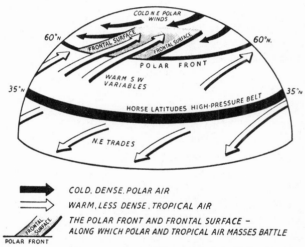

Fig. 53.—*The Site of the Battle-ground of the Polar Front*
(Northern Hemisphere)

Where the Battle Begins.—The preliminary skirmishes generally take place approximately between 45° and 60° over the oceans, where the two air forces blow more constantly than over land-masses. For this reason continental interiors are much less often affected by depressions.

Since tropical air, in contrast to the vicious onslaughts of its opponents, attacks in a somewhat mild manner, it might be

118

imagined that the battle would be rather one-sided, with the polar "heavy artillery" forcing the lighter tropical forces to retreat. The tropical air, however, by no means lacks vigour and once the battle has started it is fought to a finish. As it takes two to make a quarrel, it is obvious that if tropical air is blowing but a lull occurs in the polar winds, or vice versa, no depression can develop.

The Battle in the Northern Hemisphere.—The narrow zone on the earth's surface along which the air-masses engage with each other is called the *polar front*. This battle-line slopes gently upwards, forming an inclined plane, or *frontal surface*, which rises polewards from the polar front at an angle of much less than 1°. Fig. 53 shows how the polar front is the ground-level boundary where the polar frontal surface meets the earth's surface and how the less dense tropical air, undercut and levered up by the dense polar air, is forced to climb aloft along this frontal surface.

The battle-front, like that between human armies, does not for long remain straight. In Fig. 54 are shown stages in the course of the battle, with the polar front gradually buckling under the strain of conflict. The ground-plan for each of the various stages is accompanied by a vertical section through the depression taken along the parts marked by a dotted line.

Into the territory occupied by its polar enemies the huge volume of tropical air thrusts forward a bulge called the *warm sector*. Under the impact of this attack the polar winds, which hold the *cold sector*, wheel in an anti-clockwise direction around their tropical invaders and seek to outflank them (Fig. 54, stage 2).

As the bulging salient in the polar front slowly grows (Fig. 54, stage 3), its boundaries are given two names. Along the boundary from C, the centre of the depression, to A warm tropical air is the invader, climbing over the cold, dense polar air. Consequently along the ground from C to A the boundary between the two air-masses is known as the *warm front*. On weather maps the symbol ▰▰▰▰▰▰▰ is always used to indicate a warm front. Swirling around the warm sector, the cold polar air is violently attacking the left flank of its tropical foes along boundary CB and is forcibly driving beneath it. The warm air is thrust upwards and rearwards, and the boundary between the two air-masses from C to B along the ground is called the *cold front*. On a weather map a cold front is distinguished by the symbol ▲▲▲▲▲▲▲▲▲. The polar attackers along the cold front are not necessarily brought there by this encircling movement of the polar forces at the warm front. A large depression often draws in air from various quarters and sometimes the rearward polar air originates from a source far away from that of the polar air-masses engaged along the warm front. Although they are allies in the battle against their tropical opponents, different polar forces in a depression may vary considerably in their respective temperatures, humidity and strength.

If a fairly taut rope is sharply shaken, ripples will pass along it. Similarly, these bulges (for normally depressions, like Shakespeare's sorrows, "come not single spies, but in battalions") move along the polar front from west to east.

The line of advance taken by the centre of the depression is called the *path* (CD in Fig. 54 stages 3 and 4). Owing to the violence of the polar attack along it, the cold front travels much more rapidly than the warm front, along which the tropical forces are making a comparatively mild assault. At length the cold overtakes the warm front, at first around C (Fig. 54, stage 4), where less ground has to be covered, while finally B catches up with A. The warm sector is thus squeezed off the ground and thrust upwards by a scissors-like movement of the two fronts. When this happens the depression, now in a vigorous middle-age, is said to be *occluded* (not *con*cluded, for the warm sector still exists above ground-level). Instead of two fronts on the ground-plan there appears a single occluded front (▲▲▲▲▲▲▲, being a combination of ●●●● and ▲▲▲▲).

Up aloft the warm sector persists awhile, trapped between the inclined planes of the warm and cold frontal surfaces (Fig. 54, stage 4). Many depressions which originate over the Atlantic are occluded by the time they reach Britain. Finally, as the continental interior is approached, the warm sector disappears entirely and the depression "fills in" and dies away.

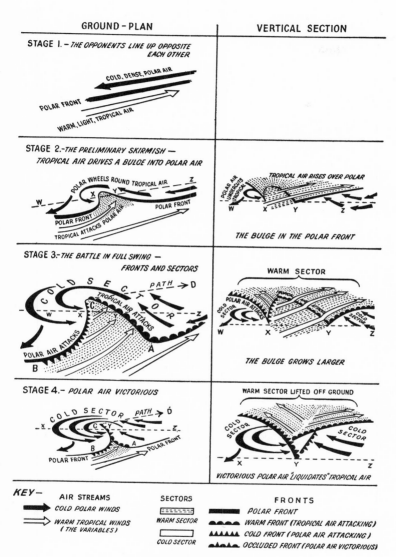

Fig. 54.—The Battle of the Depression

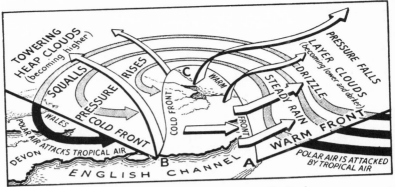

Fig. 55.—A Depression passing over England

Tropical air is indicated by white arrows, polar air by black arrows. Along the warm front (AC) tropical air rides over polar air but is violently undercut by it along the cold front (CB). Note the changes in pressure, cloud, and precipitation.

WEATHER IN A DEPRESSION

Warm Front Weather.—In the van of a depression (Fig. 55) tropical air, as it climbs forwards and upwards over polar air, is gradually chilled until condensation begins. Long horizontal sheets of layer clouds take shape, following on one another's heels in a definite order.

Heralding the approach of the depression are white wisps, several hundred miles ahead of the centre and therefore high up, for the tropical air has by now climbed to about six miles above the ground. These "mare's tails" consist of ice particles. Gradually a white, milky veil is drawn over the sky and a halo encircles the moon, a sign of rain to come. The white ice-clouds give place to light-grey water-clouds, not quite so high. Through them the sun peers as if through frosted glass, giving a greasy "watery" patch of light. A fine drizzle falls. The cloud blanket becomes lower, thicker, and darker; the drizzle passes into hours of steady rain, and visibility deteriorates. All this while the barometer has been recording a continuous fall in pressure.

Cold Front Weather.—If the depression is not occluded, an observer on the equatorial side of the path of the depression is, for a short time after the departure of the warm front, bathed

122

in tropical air at ground-level as the warm sector flows over him (∠ ACB, Fig. 55). Pressure stops falling and becomes steady.

Hurrying in rapid pursuit of the warm front, however, comes the cold front, and as it reaches him the observer notices a sharp fall in temperature and an equally sharp rise in pressure.

The polar winds are now making a flank attack on their tropical opponents, and, closing with them, are violently "shovelling" them upwards and backwards to great heights (Fig. 55).

Giant heap clouds billow upwards, resembling towers rather than the layers of the warm front. Winds reach gale force, with squalls of rain, hail, or (in winter) snow. Occasionally thunder peals and lightning flashes. These "clearing showers" are soon over. Visibility rapidly improves, patches of blue sky appear, and the depression passes away.

In an occluded depression the quiet, warm interval provided by the warm sector is missing, for the cold front has overtaken the warm front. The warm sector has been completely lifted off the ground and is now high over the observer's head.

Occasionally the battle ends very suddenly in a *line-squall*. Particularly violent gusts of wind accompany a long bank of black clouds, sharply outlined against a brilliantly clear blue sky beyond.

THE END OF THE WAR

IN an aerial war there are often several battles, or depressions. Each one is separated from its successor by a short lull of fine weather known as a ridge, or wedge, of high pressure.

From the high-pressure heart of a ridge air flows outwards on either side to fill up the low-pressure troughs of (a) the departing depression and (b) the advancing one. In fact, a ridge of high pressure really marks where the cold front of a receding depression bends round backwards to merge into the warm front of its successor. Thus, like a watershed between two rivers, a ridge is a divide between two depressions (Fig. 56). Its extraordinarily good visibility is a reliable portent of rain to come.

Every time polar winds swirl round tropical air they are brought nearer to the Horse Latitudes, so that each new depres-

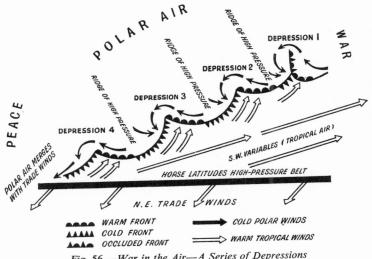

Fig. 56.—*War in the Air—A Series of Depressions*

sion is born slightly nearer to the Tropics than its predecessor (Fig. 56). Eventually, penetrating right through the lines of their tropical adversaries, the polar forces merge with the trade winds, so giving continuous air streams from Pole to Doldrums. The war is over, and a long, high-pressure anticyclonic peace prevails until the westerly variables once again muster sufficient strength to declare another war on their powerful polar opponents.

Allowing about two or three days for the passage of each depression and its succeeding ridge of high pressure, unsettled weather may last in this way for several weeks, with depression following depression until anticyclonic weather sets in for a week or two. Such rapid changes are characteristic of British weather.

The Peace of the Anticyclone

An anticyclone is a high-pressure system from whose centre winds blow outwards clockwise in the northern and anti-clockwise in the southern hemisphere (Fig. 40). Its weather, although usually settled, is not necessarily bright and sunny. Anticyclones, it is true, bring summer heat waves, but they are also responsible for winter cold waves and November fogs, as well as for "anticyclonic drizzle" and "anticyclonic gloom".

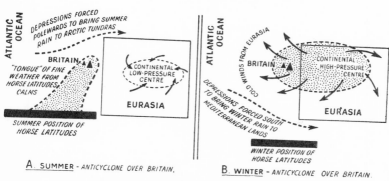

Fig. 57.—*Heat Waves and Cold Waves*

The Summer Heat Wave.—In summer the Horse Latitudes high-pressure belt swings polewards with the sun, and most summer anticyclones are "tongues" of high pressure which from the main body of this fine-weather zone push out well towards arctic regions (Fig. 57A). Those that reach Britain usually come from the Azores.

Morning dawns with the ground shrouded in mist and bespangled with heavy dew. As the sun climbs upwards the air is warmed and the mist disperses. With few clouds to interfere with the passage of heat from sun to earth, temperatures rise rapidly. Convection currents, set in motion by increasing heat, create billowing white clouds, but no rain falls, for waterdrops cannot make their way to earth through such powerful uprushes of air. As the sun sinks the air sinks likewise, for the convection currents die away and the clouds vanish. A red sunset promises well for the next day. During the cloudless, star-lit night a rapidly cooling earth gives rise to mists and heavy dew. Stagnant pools of mist are particularly thick in valley-bottoms, into which cold air drains from surrounding hillsides.

Day after day this fine, hot weather persists. Within a week or so, however, the convection currents grow so powerful that towering thunder-clouds may develop, and the fine spell breaks in a storm, rivalling those experienced practically every afternoon in equatorial regions. Following the thunderstorm, the heat wave either continues or else gives way to depressions,

125

which during the reign of the anticyclone are forced to travel farther polewards than usual (Fig. 57A).

The Winter Cold Wave.—In winter the Horse Latitudes high-pressure belt retreats in the wake of the migrating sun. In command of each of the northern land-masses is a huge high-pressure system, set up by continental cold.

Occasionally, especially in January and February, this anticyclone swells outwards to envelop the margins of continents. The British Isles, for example, sometimes receive from Russia cold, snow-laden, easterly winds which reduce temperatures to well below freezing-point. Depressions from the ocean are held at bay. Finding their progress to the north-east blocked by the barricades set up by the continental and polar high-pressure systems, they travel well southwards of these obstacles to bring winter rain to lands with a Mediterranean climate (Fig. 57B). Once it is well set, a winter cold wave, like the summer heat wave, persists for a long time. Weakening at length, it withdraws into the continent as depressions batter against it and restore mild, wet weather.

The Autumn Fog.—We "pour oil upon troubled waters", for oil, being lighter than water, rests upon it and stills its movements. Similarly, in anticyclonic calms cold, dense surface air becomes stagnant when it is trapped beneath a "lid" of warmer, lighter air floating several hundred feet above the ground. Such conditions are known as an *inversion of temperature*, for instead of falling with increasing height the temperature rises. For the cold air near the surface an escape upwards through the trap of warmer air above is impossible.

After sunset this cold ground air grows even colder and, as in the summer anticyclone, mists form. The sun, which on a summer day soon rises high enough in the sky to provide the ground with sufficient heat to evaporate the water droplets, is in autumn much lower in the sky. At first it may be sufficiently powerful to lighten the blanket of fog by mid-afternoon, when as a red disc it peers through the misty veil still clinging to the ground. Day by day, however, the fog grows thicker. The sun's

[*Manchester Evening News*
An Inversion of Temperature— "Smog" in Manchester
This photograph was taken during a morning in May. "Smog", mists (see frontispiece), fogs, and anticyclonic gloom are caused by inversions of temperature.

noon-day rays, reaching the top layers of fog at a lower and lower angle, become less and less capable of dispersing it. Although the fog may "lift" slightly during the afternoon, it "comes down" again towards sunset.

Over industrial areas, where soot mingles with the water droplets, smoke fogs pollute the atmosphere more than those of the countryside or of the sea. In fact, "smog," i.e. smoke fog, caused by an inversion of temperature, may at any season turn day into night over industrial cities.

WAR CORRESPONDENTS

IN this war of the air-masses *meteorologists*, or weather experts, act as war correspondents. They pass on news about the progress of the aerial battles to the farmer, sailor, and airman, to all of whom weather information is vitally important. They interpret to the general public the ever-changing pattern of wind and rain, cloud and sunshine, heat and cold.

Over all civilised countries spreads a close network of meteorological stations, at which meteorologists take frequent and regular observations of temperatures, winds, pressure, rainfall, state of the weather, type and amount of cloud, etc. In Europe there are over 5,000 meteorological stations, including 200 in Britain. Weather over the oceans is observed by ships, especially by weather-recording ships permanently stationed on the Atlantic. Conditions in the upper air are of the utmost

127

[*Pictorial Press*

Radio-sonde Methods obtain Upper-air Weather Information
The radio-sonde apparatus consists of a balloon carrying a small radio transmitter (at right), which broadcasts signals in tones that vary in pitch as temperature, humidity, and pressure change. Readings taken in Greenland, as shown here, are very important in weather-forecasting, since many depressions originate in this area.

importance in weather-forecasting. They are registered by aircraft and sometimes by instruments attached to a small balloon, the readings being automatically transmitted to earth by a tiny wireless set.

All this news about the weather is sent in code by teleprinter or by wireless to the headquarters of the meteorological organisation of each country. A weather map can then be drawn, a careful examination of which enables the expert to issue forecasts to cover the various regions of the country in question.

EXERCISES

1. "When two Englishmen meet their first talk is about the weather" (Dr. Johnson). "Britain has no climate, but only weather." Discuss these two statements.

2. At different seasons keep a daily weather record for about a month. Take observations at least twice daily, e.g. at 9 a.m. and 3 p.m., and record as many as possible of the following items: temperature; pressure; wind direction; type of clouds and direction of their drift; kind of precipitation (drizzle, steady rain, squally showers, dew, hoar frost, hail, thunderstorm, or snow); and visibility (record fog or mist). From your readings try to identify the passage of depressions and ridges of high pressure and also the various types of anticyclonic weather.

3. A thickening fog is said to "come down". By explaining how fog grows from the ground upwards, show how a false idea of what usually happens is given by this description.

CHAPTER XIV

The Movements of the Waters

OCEAN CURRENTS AND THEIR CAUSES

DIFFERENCES in both the salinity, or saltness, of the seas and oceans and in their temperatures bring about differences in density, which create currents. Prevailing winds, however, aided by the rotation of the earth, are by far the most important cause of ocean currents.

The rectangles A and B in Fig. 58 represent sets of continents bordering an ocean, over whose surface prevailing winds set in motion two huge "water-wheels", one rotating clockwise in the northern and the other anti-clockwise in the southern hemisphere. In the central calms of the North Atlantic swirl sargassum, a kind of sea-weed, collects to give rise to the famous Sargasso Sea, stories of which are usually exaggerated. It is no hindrance to navigation.

Various parts of these great swirls are named, usually after an important section of the coastlands along which they flow.

The currents are quite shallow troughs of water moving over the surface of the oceans. Whether they are warm or cold, compared with the oceanic waters over which they slide, depends upon the latitudes from which they originate. Warm currents, carrying tropical heat with them, move polewards (shown by full lines in Fig. 58), whereas those that flow towards the Equator are cold compared with the waters into which they are travelling (dotted lines, Fig. 58).

Bearing all these points in mind, we can now study the various currents that together make up the oceanic circulations.

To the north of the Equator the north-east trades propel westwards a warm current (2, Fig. 58), while a similar one (3) is set in motion in the southern hemisphere by the south-east trades. As these currents carry so much water away from continents B, a counter-current (1) flows back along the Equator itself to restore the balance.

On reaching the eastern shores of continents A, currents 2 and 3 are turned polewards as warm currents 4 and 5 respectively. The westerly variables, aided by the rotation of the earth, at length deflect them, current 4 to the right and 5 to the left in accordance with Ferrel's Law, so that they swing away towards continents B. Widening out and slowing down, they cross the oceans as warm drifts 6 and 7. Arriving off the western coasts of B, drift 6 is divided. Some of it continues polewards as warm drift 6(a), while the remainder flows towards the Equator (cold current 8) and finally merges with 2 to complete the northern clockwise circulation.

The southern continents do not greatly impede the easterly progress of drift 7, which therefore encircles the whole world. Some of its waters, however, are deflected by continents B towards the Equator (cold current 9), eventually to merge with 3 and so complete the southern anti-clockwise circulation.

Currents 8 and 9 become even colder when they flow alongside the hot deserts, for here prevailing off-shore trades blow the surface waters away from the coasts, thus allowing icy water to well up from the ocean deeps.

Warm drift 6(a) reaches well beyond the Arctic Circle, keeping ports such as Murmansk in Russia free from winter ice. Obviously, to balance this constant inflow of warm water into the Arctic Ocean, somewhere there must be an outflow of cold polar water. A very cold current (10) therefore moves southwards along the eastern seaboard of continents A.

In the southern hemisphere only South America extends far enough polewards to offer a passageway for a similar cold current. Here 11 flows northwards past the Falkland Islands as a southern counterpart to 10.

In every ocean these huge "water-wheels" turn, except in the northern portion of the Indian Ocean, where the powerful monsoon winds establish seasonal currents. Summer on-shore south-west monsoons push water towards India, but in winter

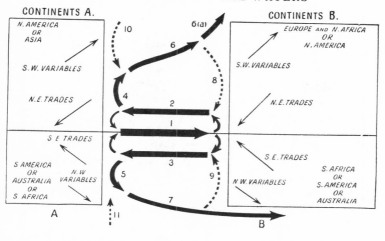

Fig. 58.—*Currents caused by Prevailing Winds and Rotation*

this current is reversed and water is propelled away from the land by off-shore north-east monsoons.

The names of the chief ocean currents are given in Fig. 58.

OCEAN CURRENTS AND CLIMATE

A CURRENT either warms or cools the prevailing winds that pass over it and, if these winds are on-shore, indirectly affects the climate of the lands washed by it. A warm current, by warming on-shore winds and so increasing their power to store water-vapour, indirectly increases both the temperature and rainfall of the land. A cold current, by chilling on-shore winds and depriving them of moisture, indirectly cools and makes dry the land towards which those winds blow.

Wherever prevailing winds are off-shore, however, the influence of a current is directed away from the land, and only a very narrow coastal belt can possibly come under its sway.

Warm currents, since they normally flow where prevailing winds are on-shore, have a more widespread effect upon the climate of the continents than do cold currents, which are usually accompanied by off-shore winds (see Fig. 24). Far removed from the oceans, continental hearts are, of course, entirely beyond the control of currents, cold or warm.

The most noteworthy effects upon climate of individual currents are as follows:

The Gulf Stream and the North Atlantic Drift.—The Gulf Stream is particularly warm and powerful, for not only does it receive water from the North Equatorial Current, but it is also reinforced by some from the South Equatorial Current, turned into the Gulf of Mexico by the wedge-shaped "shoulder" of North-east Brazil (Fig. 59). At first it merits the name "stream", for it is narrow and fast, flowing at six miles an hour, and is clearly distinguishable by its deep-blue colour. Gradually, however, it slows down and widens until, as the North Atlantic Drift, it becomes a sluggish crawl of the whole

Fig. 59.—How Brazil helps to make British Winters very Mild

surface waters, blown towards Western Europe by the prevailing winds. Every minute it is said to bring with it as much heat as would be generated by burning 2,000,000 tons of coal.

In winter the north-western margins of Europe are enclosed in an aerial *gulf of winter warmth*, created by prevailing southwest variables blowing over this warm drift. From the Atlantic this gulf stretches north-eastwards over Britain to the shores of Norway. Regions lying within it enjoy mean January temperatures between 30° F. and 40° F. above the average for their latitudes.

In the absence, except in Norway and Sweden, of high north-south mountain barriers to hinder the progress of these on-shore winds, winter warmth from the current penetrates well into Western Europe.

By contrast, over the heart of Eurasia, far beyond the sway of the sea, a *bay of winter cold* reaches southwards well towards the tropics. Both the gulf of winter warmth and the bay of winter cold are well shown by the course of the 32° F. (freezing-point) isotherm for January. Over Norwegian coastlands it pushes north-eastwards to well inside the Arctic Circle before sagging southwards to about 40° N. over Central Asia. Although it enters North-west Europe at a point lying nearly 3,000 miles north of the Tropic of Cancer, it leaves Eastern Asia within only 800 miles of the Tropic, at about 35° N. Here, of course, the warm Kuro Siwo is flowing away from the land, while cold winds blow outwards from the high-pressure continental "lungs" (Fig. 60).

The Kuro Siwo and the North Pacific Drift.—Although January

Fig. 60.—*Gulfs of Winter Warmth and Bays of Winter Cold*

+ or — degrees Fahrenheit shows by how much mean January temperature rises above or falls below the average for the latitude.

[Camera Press

A Gulf of Winter Warmth

In this mid-winter scene at Seward, Alaska, snow is seen lying on the mountains but the sea is not frozen because of the effect of the warm North Pacific Drift and of on-shore variables.

temperatures in the temperate maritime coastlands of British Columbia, Washington, and Oregon are forced up to between 10° F. and 20° F. above the average for their latitudes, the effect upon climate of these warm currents and the on-shore variables accompanying them is less striking than that of the Gulf Stream and the North Atlantic Drift. To begin with, these currents are less powerful than their Atlantic counterparts; moreover, high north-south mountain barriers restrict their influence to the coasts.

Over North America the 32° F. isotherm for January repeats its Eurasian pattern. In fact, it drapes itself festoon-wise across the map of the northern hemisphere. It pushes polewards over western margins in gulfs of winter warmth, but sags equatorwards in bays of winter cold over continental interiors and eastern margins, where temperatures sink in North America to 20° and in Asia to 30° below the average for their latitudes (Fig. 60).

In the southern hemisphere the absence of large land-masses prevents the development of gulfs of winter warmth and bays of winter cold, although the West Wind Drift certainly exerts its mild influence over the winters of South Chile, Tasmania, and the South Island of New Zealand.

Icebergs off Newfoundland [*Royal Canadian Navy*

The Labrador Current carries icebergs southwards from the Greenland ice-cap. By early summer they reach shipping routes across the Atlantic. Since the loss of the liner *Titanic* in 1912 patrol boats, operated by the United States Coastguard Service but maintained by all nations whose ships navigate these seas, have transmitted wireless reports about the movements of icebergs.

Canaries, Californian, Humboldt, Benguella, and West Australian Currents.—All these cold currents wash western coasts where for much of the year, and in hot deserts for the whole year, there are off-shore trades. Their control over climate is therefore limited to coastlands, but is there very remarkable. In 1948, during an August heat wave along the Atlantic coast of the United States, municipal workers were granted a half-holiday when the temperature soared almost to 100°F. On the same day on the shores of the Pacific Ocean, San Francisco's postmen, clad in winter overcoats, groped through a blanket of fog created by the Californian Current. With a mean July tempera-ture of 57° F., San Francisco is cooler and foggier than London (July 64° F.), although it is almost 1,000 miles nearer to the Equator. Only 70 miles inland, however, lies Sacramento, whose July temperature of 73° F. shows that the current's control is coastal only.

Similarly, coasts washed by the Benguella Current experience temperatures 10° below the average for their latitudes, while the Humboldt Current brings to Peruvian coasts chilly sea-fogs which for days blot out the sun.

Labrador and Oya Siwo Currents.—These currents contribute to the coldness of winter along the north-east coastlands of North America and Asia. In spring, when it carries icebergs from the melting fringe of the polar ice-cap, the Labrador Current is particularly cold. Off the coasts of Newfoundland condensation in the form of fog results when cold air above these icy waters mingles with warm air above the North Atlantic Drift (page 106). The combination of fog and icebergs makes great circle voyages from America to Europe very dangerous, and in the critical period of spring and early summer vessels often take a longer but less perilous course.

Similar fogs occur off North-east Asia, where the cold Oya Siwo Current approaches the warm North Pacific Drift.

EXERCISES

1. Liverpool and Bristol prospered on profits made on the "triangular run" taken by slave-traders:

Voyage	*Trade*
1. Britain to W. Africa	Manufactured goods exchanged for slaves.
2. W. Africa to New World	Slaves for tobacco, sugar-cane, rum.
3. New World to Britain	The above "semi-luxuries" sold at good profits.

What prevailing winds and ocean currents helped the ships along each side of this triangular run? Give examples to show how long sea-routes are still partly determined by winds and currents.

2.

Place	*Mileage from Equator*	*Position in Eurasia*	*Lowest Mean Temperature*
Deerness (Orkneys)	4,100 miles	Western Margins	38° F.
Baku (U.S.S.R.)	2,800 ,,	Interior	38° F.
Shanghai (China)	2,200 ,,	Eastern Margins	38° F.

Explain why Deerness, Baku, and Shanghai, although greatly differing in their distance from the Equator, have exactly the same lowest mean temperature in winter.

3. Why is coral found on the tropical coasts of East Africa and East Australia and not on coasts in corresponding latitudes on the west of those continents?

PART II

NATURE'S RESPONSE TO CLIMATE—NATURAL VEGETATION

CHAPTER XV

The Earth's Carpet of Vegetation

VEGETATION AND ANIMALS

DIRECTLY or indirectly, all animals depend upon plants. Thus grass feeds the *herbivorous*, or plant-eating, antelope which in its turn is preyed upon by the *carnivorous*, or flesh-eating, lion. Different kinds of *flora*, as the natural vegetation of a region is called, house different kinds of *fauna*, or animal life. Particular animals are not necessarily limited to particular regions, yet rarely do they wander far from their natural surroundings. For instance, the camel does not plough through polar snows, neither does the reindeer browse in swampy equatorial forests; the giraffe does not stray beyond its tropical savanna home, nor the llama from its Andean pastures.

CLIMATE, VEGETATION, AND MAN

MAN lives upon both plants and animals. In many parts of the world he satisfies his needs with his own hands by growing food-crops, by herding or hunting animals, by fishing, or by gathering fruits. Elsewhere he may be a miner, a merchant, or a factory-worker, but he depends none the less upon plants and animals for food, clothing, and the furnishings of his home. Even the tools and machines that men use depend indirectly upon energy derived from plants. "The plough, the spade, and the cart must eat through man's stomach; the fuel that sets them going must burn in the furnace of man or of horses. Man must consume bread and meat or he cannot dig; the bread and meat are the fuel which drive the spade. . . . Without this fuel,

[*Black Star*

Rabbits in Australia cause Soil Erosion

White settlers, by introducing rabbits into Australia, have interfered disastrously with the *balance of Nature*. With no natural enemies to keep down their numbers, they became a pest. They foul pastures and destroy grass by close grazing, so encouraging soil erosion. Farmers have lost millions of pounds, for five rabbits eat as much as one sheep. Scientists hope to check the pest, for some rabbits have been injected with myxomatosis, a microscopic virus which causes fever, paralysis, and death. This disease will spread—unless the rabbits build up an immunity to it.

the work would cease as an engine would stop if its furnaces were to go out." (Samuel Butler.)

Although man is not so firmly attached as animals are to a natural homeland, and although he is able to roam over the whole world, he usually settles in those regions in which he feels most comfortable. Thus the Eskimo does not seek to exchange his tundra surroundings for the equatorial forest home of the pygmy, while the Australian aborigine would soon die in arctic snows.

Of all the races of mankind those from temperate latitudes, and especially from Western Europe, have wandered farthest from their homelands. The climate of tropical jungles and deserts saps man's energy, whereas in arctic tundras man alternates between a summer spurt of energy to store up food and a long rest in winter, when lack of daylight and icy snows bring his outdoor activities practically to a standstill. In tem-

[*Mondiale*

The Dust-bowl of America

In many parts of the world fertile land has been converted into desert by reckless farming methods.

perate lands, however, man is kept vigorous and alert, both physically and mentally, by the ever-changing weather brought about by alternating depressions and anticyclones. He can work steadily throughout the year, yet he finds sufficient leisure to cultivate the powers of his mind. In such surroundings the arts and sciences flourish and civilisation progresses apace.

The fact that the welfare of the human race largely depends upon the plant growth that covers the earth becomes evident if we consider the disasters which have overtaken man when, in trying to become master of his surroundings, he has torn up the soil's carpet of vegetation.

In some regions man has ploughed up grasslands or has pastured too many animals upon them; in others he has ruthlessly cut down forests for timber or to make room for his crops. In each case the soil has been robbed of the vegetation cover that for hundreds of years had held it firmly in place. Heavy rains have then washed away the loosened soil, which has clogged the rivers and so has caused disastrous floods. In times of drought the soil, powdered under a hot sun into fine dust,

Contour Ploughing

Ploughing along slopes instead of up and down them is now practised in hilly country. How does this method check soil erosion?

has blown away. True, in many lands man has, by careful irrigation, made the desert "blossom as the rose", but far too often has he reversed the process and turned fertile regions into man-made deserts like the famous "dust-bowl" of the United States. Ignoring Sir Francis Bacon's wise advice that to tame Nature man must first understand her, he has paid a heavy

[C. J. Martin

Root System of a Plant in Somaliland
In hot deserts long roots support a stunted surface growth. Compare the size
of the roots with the man's height.

penalty for his thoughtlessness. It takes several hundred years
to form an inch or two of soil, which by this process of *soil
erosion* can be blown or washed away in a few days, or even in
a single tropical storm. In the United States alone an area of
three times that of the total farmlands of Britain has been
ruined. With its population increasing by 80,000 a day, the
world can ill afford such reckless "mining" of its soil.

How Plants Feed

A PLANT, like an animal, must feed. From the soil its roots draw
up water in which are dissolved life-giving salts. The plant also

141

takes from this water hydrogen, and forms food by combining it with carbon from carbon dioxide drawn in from the air. This indispensable soil moisture has, of course, been received by the earth as rain, without which plants would therefore die. Once it has satisfied its needs, a plant gets rid of its surplus water through its leaves. A tree, by this process of *transpiration*, may give back to the atmosphere some eighty gallons of water a day.

Some plants need more water than others. Trees require most, while in areas which suffer from droughts plants are especially equipped by Nature to make a little moisture go a long way. Plants and animals which are fitted to overcome handicaps of their homeland such as excessive rain or drought, heat or cold, are said to be *adapted to their environment*. To adapt is to make fit, while environment means surroundings.

How Plants Fight Drought

PLANTS that are adapted to drought protect themselves in three ways:

(*a*) *By obtaining extra water-supplies.*—In dry lands the roots of plants spread out in a vast underground network in search of water. Some desert plants support a growth of one or two feet above the surface upon an immense foundation of roots which either penetrate the ground to great depths or stretch far outwards just below it.

(*b*) *By storing water.*—Moisture is stored in bulbous roots, fleshy leaves, and swollen trunks. In tropical savannas the trunk of the baobab tree serves as a natural water-butt, while thirsty Australian aborigines can draw about a quart of water from the root of the mallee plant.

(*c*) *By preventing loss of water.*—Having obtained and stored their water supplies, plants continue their struggle against drought by checking the loss of water caused by transpiration. In hot, dry weather the rate of transpiration rapidly increases and a plant must somehow defend itself against the hot sun. Sometimes the leaf is protected by a hairy or waxy covering; sometimes it is glossy so that sunlight is reflected. To present as little surface as possible as a target for the sun's rays, leaves may be small, or thorny, or hang vertically. They may possess

A Baobab Tree

[*Popper*

By casting its leaves and storing water in its huge trunk, the baobab maintains life through the dry season of savanna lands.

few of the pores through which moisture is lost. Some plants manage with few leaves or even none at all. A thick, protective bark is yet another way of reducing loss of water.

Plants differ in the extent to which they are adapted to withstand drought. For those growing in climates that are dry for only a part of the year the fight to get and to keep water is strenuous, but for those in absolutely rainless deserts the endless struggle against constant drought is a matter of life and death.

PLANTS AND "COLD-WEATHER DROUGHT"

IN some regions plants, although not faced with drought, may labour under as great a disadvantage as if they were. Long periods of cold, or even cool weather with occasional frosts, prevent the food-bearing water from circulating freely through plants. In fact, whenever mean monthly temperatures fall below 42° F. plant-growth normally stops.

During this winter resting-season, when little or no water can be taken up from a frost-bound soil, it would be fatal if the water supply already held by a plant were diminished. Transpiration must therefore be stopped by the shedding of leaves. Plants which lose their foliage in order to rest through a drought, not necessarily in winter but at any season, are called *deciduous*.

By contrast, *evergreen* plants, basking in warm winters, are not faced with any such problem. They retain their leaves and continue to grow throughout the year. Cone-bearing, or *coniferous*, trees like the pine and fir, however, remain evergreen even where winters are severe and the ground is frozen. They can afford to keep their leaves because:

(*a*) Their need of water is small and easily satisfied, compared with that of other trees.

(*b*) Their leaves are small and needle-shaped, and the drain upon their water supply by transpiration is therefore slight. Moreover, snow easily slides off such leaves and the boughs are not broken by its weight.

(*c*) Wherever winters are very long as well as cold trees cannot spare the time to produce new foliage in spring. From the very start they must be fully equipped with leaves in order to take immediate advantage of the short summer growing-season.

Coniferous forests consist of softwoods, whereas broad-leaved evergreen and deciduous forests contain mainly hardwoods.

Like those in hot deserts, plants in *tundras*, or arctic wastes, are small and stunted. In hot deserts roots must go far underground to maintain a little growth on top. In these arctic deserts roots cannot penetrate far enough to support more than a dwarfish growth above the surface, for winter frosts harden the ground, while in the short summer the sun thaws no more than a thin top layer of soil.

EXERCISES

1. Distinguish between ordinary drought and "cold-weather drought". How do plants survive in each of these kinds of drought?

2. "In remote times men saw trees as gods, as guardians of the fertility of the soil. Our destruction of those guardians has sacrificed fertility with all the horror of dust storm, drought and flood." Explain this statement.

3. "The prosperity of New Zealand is based on grass." Explain this statement.

CHAPTER XVI

How Climate Helps to Control Natural Vegetation

SINCE forest, grassland, and desert vegetations differ greatly in their climatic requirements, a clue to the kind of climate to be expected in any given region can often be provided by the natural vegetation that covers the area.

FORESTS

Rainfall and Forests.—Trees demand more water than other plants. In temperate regions a moderate rainfall will support forest-growth, but in hot lands trees need a heavy rainfall to make up for the great loss of water by rapid evaporation. Resisting the drought by shedding their leaves, deciduous trees may flourish in regions with a dry season, but instead of forming dense forests they either grow as sparse woodlands or are scattered as isolated specimens over a park-like landscape, as in tropical savannas. For a region to be thickly carpeted with forest, however, the rainfall must be well distributed throughout the year. Using forest as a "signpost" to climate, we can usually assume that wherever it grows it indicates that:

(*a*) Rain falls at all seasons.

(*b*) In temperate lands there is at least a moderate annual rainfall of from 20″ to 40″.

(*c*) In hot lands the rainfall is at least heavy (40″ to 60″) and is usually very heavy, with over 60″.

In geography few rules are without exceptions, and to those given above four such exceptions are outstanding:

1. Theoretically, monsoon lands, with their long winter drought, should not be forested. Nevertheless, certain parts are so heavily drenched by summer storms that enough moisture soaks into the swampy soil to supply trees throughout the dry

[Black Star

A Temple in Siam overgrown by Tropical Rain Forest

This picture shows how rapid and powerful plant-growth is when heat is combined with heavy rain.

season. Most of these trees are deciduous, shedding their leaves to check transpiration during the drought.

2. In cold lands a light annual precipitation is enough to support coniferous forests, partly because they need less water than other trees, and partly because little water is lost through evaporation from the ground or through transpiration from the needle-shaped leaves.

3. In lands with a dry season long and narrow *gallery forests* border watercourses, for here river-water rather than rain sustains the trees.

4. In some areas the character of the soil affects the vegetation cover, which shows that while climate is the chief factor it is not the only factor to be taken into consideration.

Temperature and Forest.—Two mean monthly temperatures help to control the growth of forests, namely 50° F. in summer and 42° F. in winter. Trees cannot grow at all unless the temperature for the warmest months reaches at least 50° F., while

146

in winter a mean monthly temperature of 42° F. halts the growth of most plants. The kind of forest that clothes a region therefore depends upon how many winter months experience temperatures of under 42° F., as shown below:

Months below 42° F.	Type of Forest	Why
None	Broad-leaved evergreen	No cold-weather drought, thus leaf-shedding to check transpiration at any one particular season is unnecessary.
1 to 6	Deciduous	Cold-weather drought occurs, thus leaf-shedding is necessary, but summer is long enough for new foliage to grow.
Over 6	Coniferous evergreen	Very long cold-weather drought, thus summer is too short for new foliage to grow. Transpiration through needle-shaped leaves is slow, thus a little moisture goes a long way.

The result of falling temperatures upon tree-growth is very well shown by the changes in vegetation to be seen as we climb mountains. Leaving the broad-leaved evergreen and deciduous forests of lower levels, we reach the conifers which clothe the colder heights, while above the *tree-line* it becomes too bleak even for these hardy trees.

Using forest as an indicator of temperate conditions, we may say that:

(*a*) Broad-leaved evergreen forests suggest winter warmth.

(*b*) Deciduous forests in temperate lands point to a cool or cold winter, but not a particularly long one, whereas in hot lands they indicate a dry season.

147

[*P. B. Redmayne*

Tropical Grassland

In tropical savannas summer heat and rains encourage the growth of tall grasses. Scattered trees grow but are adapted to resist drought in the cooler season, when the grass withers. A scene in the north of the Gold Coast savannas of West Africa.

(*c*) Coniferous forests usually hint at a very long, cold, and snowy winter. These "Christmas trees", however, are not always reliable indicators of severe winters, for they are not confined to cold lands. For instance, pine trees, provided that the soil is sandy, grow to within a few degrees of the Tropics.

GRASSLANDS

Rainfall and Grassland.—Few people in Britain would mow a lawn at Christmas-time, for in winter grass ceases to grow except in a few sheltered areas in the south-west. After this winter rest it springs up with renewed vigour and grows apace during the summer, when it demands not only warmth but also rain to encourage its growth. Grass therefore thrives in lands of summer rain and winter drought, a seasonal rhythm which admirably suits its alternating periods of growth and rest.

Wherever grassland is the natural vegetation it points to the following facts about the climate:

(*a*) Most of the rain falls in summer.

(*b*) The annual rainfall is considerably less than that of

Desert Vegetation   [*Popper*
In hot deserts vegetation is sparse. Through which canal is the ship passing?

forested regions. In temperate lands it will be light, with from
10″ to 20″ or thereabouts.

(*c*) In hot lands, where evaporation is rapid, there will
probably be a moderate rainfall of approximately 20″ to 40″.

Temperature and Grassland.—Wherever intense heat accom-
panies summer rain the grass shoots up to a height of several
feet. In tropical *savannas* scattered, drought-resistant trees
break the monotonous landscape of tall grasses, while gallery
forests fringe the streams. To use it correctly, the word *savanna*
should be reserved as a name for *treeless plains*. It has, however,
been widely accepted as relating to tropical parklands, although
a more accurate description for this kind of vegetation is
grass with woodlands. In temperate grasslands, e.g. *prairies*
and *steppes*, summers are less hot than in savanna lands.
Consequently the grass here, especially in steppe-lands, is
shorter than it is within the Tropics, and there are few trees
apart from riverside gallery forests. The term *steppe* is also
applied to sparse grassland in tropical lands.

DESERTS

Rainfall and Deserts.—Regions with a very light rainfall of
under 10″ are classed as deserts. Plants that grow here are very
well equipped by Nature to fight against persistent drought in
order to get and to store enough water for their needs.

Temperature and Deserts.—Climatically there are three main types of desert:

(*a*) Hot deserts lie on the western margins of continents in latitudes where the trade winds are at all seasons off-shore.

(*b*) Far from the oceans, temperate deserts and semi-deserts occur in the driest parts of the interior of North America and Asia. In the southern hemisphere Patagonia, in Southern Argentina, is a temperate semi-desert within the rain-shadow area of the Andes.

(*c*) Across the far north of Eurasia and North America stretch cold deserts, where tundra vegetation springs from the thin film of soil which thaws in summer. Tundra vegetation also clothes the higher slopes of mountains, immediately below their snowy peaks.

Between natural vegetation regions, as between the climatic regions which they so closely match, there are no definite boundaries. Forest gradually merges into grassland, and grassland fades slowly into desert (Fig. 61).

CLIMATIC REGIONS

IN the remaining chapters we shall look at plant, animal, and human life in the various climatic regions of the world. As we

Temperature Belt	Sub-divisions	Alternative Names
Very cold	——	——
Cold	——	——
Cool Temperate	1. Temperate Maritime	Western Cool Temperate, Oceanic, or British.
	2. Cool Temperate Continental	Interior Cool Temperate.
	3. Laurentian	Eastern Cool Temperate.
Warm Temperate	1, Mediterranean	Western Warm Temperate.
	2. Warm Temperate Continental	Interior Warm Temperate.
	3. China	Eastern Warm Temperate.
Hot	1. Monsoon	——
	2. Hot Desert	——
	3. Tropical Maritime	——
	4. Tropical Continental	Sudanese.
	5. Equatorial	——

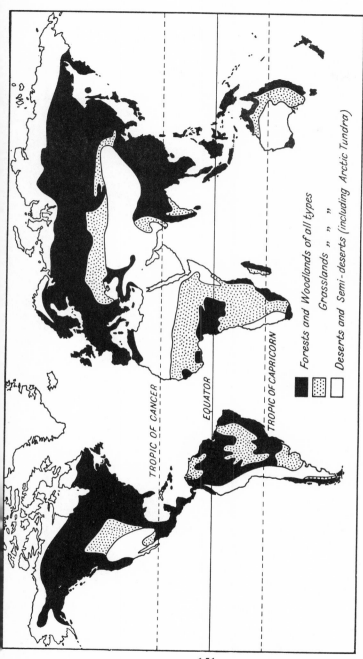

Fig. 61.—The Distribution of Natural Vegetation

Forests and Woodlands of all types

Grasslands " " "

Deserts and Semi-deserts (including Arctic Tundra)

Note: 1. In equatorial regions and eastern margins forests are practically continuous.
      2. Grasslands are mainly in continental interiors and often form a transitional "stepping-stone" between forests and deserts.
      3. Hot deserts are along western margins, apart from the great desert belt from North Africa into Central Asia.

[*E. O. Hoppé*

The Fight against Tropical Diseases

The *Anopheles* mosquito carries the malaria parasite, while the *Aedes* mosquito similarly spreads yellow fever. By destroying mosquitoes these diseases are held in check. (Left) Taking a sample, for laboratory tests, of mosquito larvæ. (Right) Killing larvæ by spraying oil over breeding-grounds.

study each region we shall notice how the climate and vegetation of man's homeland greatly influence his activities.

Climatically the world can be divided into five huge temperature belts, three of which are sub-divided into smaller regions, as shown in the table at the bottom of page 150.

In regions with an unfavourable climate civilised man has brought science to his aid in overcoming the various obstacles that bar his progress. In cold lands he dwells in comfort in a house warmed by central heating. To some extent he triumphs even over hot, swampy, equatorial forests by fighting insect pests and by conquering tropical diseases. By special methods of farming and by planting drought-resistant seeds he successfully cultivates very dry lands, while even arctic regions can be made to produce food.

The more primitive man is, of course, the more complete is Nature's command over him, but the more civilised he becomes, the more easily does he master his surroundings. Civilisation is, in fact, largely the result of man's eternal struggle with Nature. Wherever this struggle for existence is too severe, however, man becomes so absorbed in getting enough food to eat that he finds little leisure to devote to music, literature, painting, and the many other arts of a life lived at its best. Unfortunately, in his haste to improve his lot man has, as we have seen, sometimes tried to master Nature by force rather than to coax her to help him, and by doing so he has often brought disaster upon himself.

EXERCISES

1

Station	J	F	M	A	M	J	J	A	S	O	N	D
A. Temp (°F.)	−34	−36	−30	−7	15	32	41	38	33	6	−16	−28
Prec. (ins.)	0·1	0·1	0	0	0·2	0·4	0·3	1·4	0·4	0·1	0·1	0·2
B. Temp.	72	71	69	64	59	54	52	55	59	63	67	70
Prec.	3·6	4·4	4·9	5·4	5·1	4·8	5·0	3·0	2·9	2·9	2·8	2·8
C. Temp.	36	39	43	50	56	62	65	64	59	51	43	37
Prec.	1·5	1·2	1·6	1·7	2·1	2·3	2·2	2·2	2·0	2·3	1·8	1·7
D. Temp.	77	81	89	94	96	91	84	82	82	84	83	77
Prec.	0	0	0	0	0·6	3·9	8·3	8·3	5·6	1·9	0·3	0·2
E. Temp.	27	27	30	38	46	53	57	56	50	41	33	27
Prec.	3·3	2·3	2·5	2·2	2·3	2·5	2·6	2·6	3·3	4·3	3·5	4·3
F. Temp.	71	71	69	65	63	62	60	61	63	64	67	69
Prec.	0	0	0	0	0	0	0·02	0	0·03	0	0	0
G. Temp.	74	75	78	82	85	82	79	79	79	81	79	76
Prec.	0·1	0	0·1	0	0·7	20·6	27·3	16·0	11·8	2·4	0·4	0
H. Temp.	7	10	23	43	54	63	69	68	58	45	27	15
Prec.	0·5	0·5	1·1	1·8	2·4	3·4	2·2	2·0	1·2	1·0	0·7	0·6

Each of the places A to H is in a different climatic region. From the temperature and precipitation figures given, describe each climate in the way shown on page 56 (exclude the cause of precipitation).

2. Each of the above climatic regions has one of the following types of natural vegetation: coniferous forest, hot desert, temperate grassland, deciduous forest, tundra, monsoon rain forest, broad-leaved evergreen forest, and tropical grassland (savanna). Identify the natural vegetation of each region from the information given about the climate.

PART III

MAN'S RESPONSE TO CLIMATE AND VEGETATION

CHAPTER XVII

The Very Cold Belt

WORLD POSITION AND CLIMATE

THE very cold belt covers those North American and Eurasian margins that fringe the Arctic Ocean. Along the western margins of these continents its southern boundary is roughly the Arctic Circle (66½° N.), but in Eastern Canada it reaches as far south

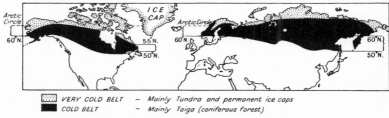

░ VERY COLD BELT	– Mainly Tundra and permanent ice caps
■ COLD BELT	– Mainly Taiga (coniferous forest)

Fig. 62.—The Cold and Very Cold Belts

as 55° N. (Fig. 62). Owing to the absence of land in similar latitudes south of the Equator, there are no southerly counterparts to these lands. Fig. 63 gives the main facts about the climate of the very cold belt. It is important to understand the diagram, for in later chapters each climate will be illustrated in this way. Notice the following points:

Temperature.—The word-scales for the temperature "ladders" and for the annual range are those given on page 55. When studying these diagrams, always remember that the words on the "ladders" refer to the state of the *air*. They should never be applied to the *temperatures* given, which, being numerical, cannot be described by terms such as hot or cold (see page 55).

154

Precipitation.—The annual amount is shown both by the word-scale given on page 55 and by the total length of the horizontal strip, drawn on a scale of 1 millimetre to 1 inch of rain. The seasonal distribution is shown by the unshaded and shaded parts of this strip, as follows:

Percentage of total precipitation falling in winter half-year is unshaded. Percentage of total precipitation falling in summer half-year is shaded.

For the northern hemisphere the winter half-year has been taken from November to April, with the summer half-year from May to October, and vice versa for the southern hemisphere. Beneath the strip are given reasons for the seasonal distribution of precipitation.

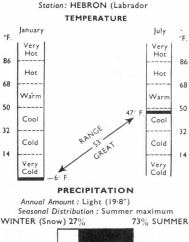

Fig. 63.

With such a diagram as a guide and using the headings given on page 56, the climate can be described as follows:

Temperature

1. The *summer* is cool.
2. The *winter* is very cold.
3. The *annual range* of temperature is great.

Precipitation

1. The *annual amount* of precipitation is light.
2. The *seasonal distribution* is uneven, for usually there is a definite maximum in summer.
3. *Form and Cause.* Much of the summer rainfall comes from depressions, which invade these arctic regions when the wind and pressure belts sway polewards with the migrating overhead sun. Moreover, in summer the air, being warmer, can hold more water-vapour than it can in winter, when cold, dry winds prevail. Such precipitation as does fall in winter takes the form of snow, for temperatures then sink to well below freezing-point.

155

Natural Vegetation and Animals

NOT even the cold-resistant conifer can live in these arctic lands, since the temperature of the warmest month fails to reach the 50° F. summer limit for tree-growth. In winter a blanket of snow covers the frozen ground. Summer comes with a rush, and the melting snow drains into hollows to form swamps. Good drainage is impossible, for the ground beneath the thin top layer is permanently frozen. Moraines, left behind after the Ice Age, provide hillocks of better-drained land. Rising above the swamps, they are carpeted with stunted mosses, lichens, berry-bearing bushes, dwarf willows, and, especially on the warmer southern slopes, with brightly coloured flowers. These treeless hummocky lowlands are called tundra, a Russian word meaning wasteland. Plants in these cold deserts shun the swampy hollows and cling to the drier hillocks. This habit is reversed in hot deserts, where plants seek water in the oases of hollows but avoid the drier areas.

Tundra swamps breed mosquitoes by the million. The reindeer, caribou, and musk-ox roam on land, while the seal, walrus, polar bear, and whale sport in arctic waters. In summer many animals from coniferous forests to the south invade the tundra in search of food, and migratory birds are attracted here by the wealth of insect life.

Man in the Very Cold Belt

The Eskimo.—Along the arctic fringes of North America and in Greenland live the Eskimos, or Innuit, as they call themselves. Life for the Eskimo is an endless struggle against a cruel climate and hostile surroundings. Often menaced by starvation, he displays a wonderful appetite whenever meat is plentiful, gorging himself in banquet after banquet, since he literally does not know where his next meal is coming from. Food is often regarded as common property and while the supply lasts nobody is allowed to go hungry. A violent death from exposure to cold or by accident when hunting or fishing is only too common. Snow-blindness constantly threatens him, and is warded off by wearing bone goggles.

Since much of his food comes from the sea, the Eskimo usu-

An Igloo

This igloo is built on sea-ice. The children are watching for their father's return from a seal hunt.

ally lives along coasts. Besides providing him with a food supply, sea-animals satisfy many of his everyday needs. They yield oil for lighting and heating, bones for weapons and sledges, sinews for thread, and skins for clothing and boat-coverings. The sea, too, brings him driftwood, so valuable in these treeless wastes.

He is not, however, entirely dependent upon the sea. In the long days of the arctic summer he wanders inland over the swamps, searching for birds' eggs and berries and killing the caribou, which supplies him not only with meat but also with skin for clothes and fat for fuel. Caribou flesh is eaten in a cooked, dried, or even raw state—in fact, the word Eskimo means "eater of raw meat". Some is frozen and buried beneath stones, safe from wandering animals, to tide him and his dogs over winter days of scarcity. Caribou fat, like seal oil, is burned to give light and heat. Berries are kept in ice, and birds' eggs are considered a great delicacy. Summer is also the season when fish are trapped or speared in river, lake, and sea. For sea-fishing the Eskimo uses a kayak. This light boat of sealskin stretched over a framework of driftwood or bone fits its owner like a glove. The umiak, a much larger boat, carries the whole

157

family. A skin tent serves as the Eskimo's "summer-house".

Autumn sees him returning to the coast to hunt the polar bear, whose skin is at this season in prime condition in preparation for a long winter sleep. Formerly the bear was chased and held at bay by the Eskimo's husky dogs until it could be speared, but nowadays the long-range rifle has replaced the short-range spear.

Later on, before ice closes over the sea, the walrus is harpooned from a kayak and, after a struggle, is brought to the surface to be shot. Hunks of walrus meat, like caribou flesh, are stored beneath stones.

Of even greater importance to the Eskimo than fish, caribou, polar bear, or walrus is the seal. Its skin, chewed by the women until it is pliable, is made into a coat, a hood, and water-proof boots. The coat is in summer discarded for a lighter one of fox fur. Garments of caribou skin are, however, worn by those tribes that live inland, and they are warmer, though less waterproof, than sealskin clothing. Beneath the caribou top-coat a shirt of birdskins, with the feathers turned inwards, ensures warmth. Both men and women wear trousers of bearskin.

In winter the Eskimo, besides hunting seals, visits his traps, in which are caught the arctic fox and other animals whose valuable furs are sold to the white trader. He travels in a sledge of wood and bone, drawn by half-wild husky dogs. Walrus meat is carried for himself and for his dog-team, and for shelter he builds a small *igloo*, or snow-house. Travel through the long arctic night is possible in the light of the moon and of the beautiful northern lights.

His trapping expeditions over, he returns to his home, an igloo big enough to shelter two families at once. It may be twelve feet high and thirty feet across. Shaped like a beehive, the igloo is made of blocks of snow. A tunnel-like entrance, in which huskies are kennelled, leads to a skin-draped interior. Here much of the long dark winter is passed, the men making weapons or preparing furs for sale, and the women sewing clothes and cooking meals over seal-oil fires. Occupying much of the floor is a skin-covered ledge of snow upon which the Eskimos sleep, cosily wrapped in caribou blankets. Sometimes a permanent hut of driftwood or of stone is built, the chinks

Eskimos in a Wooden Hut
Some Eskimos live in wooden huts. Note the manufactured goods which they can get in exchange for furs.

being cemented with earth. Indeed, some Eskimos have never seen a snow igloo.

At length winter ends, and with summer's return preparations are again made to travel inland to hunt caribou.

To some extent civilisation has reached Canadian tundras. The Royal North-West Mounted Police maintain law and order. Sometimes the "mountie" covers hundreds of miles of arctic wastes to "get his man". His duties are not, however, confined to arctic lands, for he fights fires in forest and prairie and throughout Canada keeps order among lumberjacks, fur-trappers, miners, and farmers. Occasionally the aeroplane invades the solitudes of the tundra to pick up a cargo of furs or to survey the country, for in uncharted lands aerial photography plays an important part in the discovery of minerals. In spite of his primitive and superstitious beliefs, the Eskimo is a born mechanic and often helps white miners and traders.

At trading-posts maintained by the famous Hudson's Bay Company the Eskimo exchanges his furs for rifles, binoculars, knives, sewing-machines, gramophones, and foodstuffs such as lard, flour, and tea. Formerly he drank the water in which his meat was boiled, but he can now brew strong black tea. After using them he stores the tea-leaves for times of scarcity. The more he needs these foodstuffs, the more is he prepared to pay in furs, and so greatly are they prized that he will not share them with his companions as he will his seal or caribou meat.

Civilisation has also brought him schools and churches, but at heart he remains superstitious. Possibly, with the Australian aborigine, the Kalahari Bushman, and other primitive peoples, he may die out as the white man in increasing numbers invades his homeland. Certainly, he often catches diseases when, in copying the white man's habits, he abandons the natural way of life to which his ancestors were accustomed.

The Lapp.—Lapland, Europe's tundra region, covers the arctic parts of Norway, Sweden, Finland, and Russia. In build and speech Lapps are unlike other European races, and they were pushed by more advanced peoples into their present bleak home.

There are three types of Laplander:

(a) *Fjeld, or Mountain, Lapps.*—In summer these people climb the mountains in search of pasture and stunted mosses for their reindeer, and in winter descend to the shelter of the valleys. This migration to and from mountain pastures is called *transhumance* and is a common practice in mountainous lands.

(b) *Coastal and Riverside Lapps.*—Unlike the Fjeld Lapps, these folk live by fishing, and instead of moving up and down the mountains they migrate northwards and southwards with the changing seasons.

(c) *Forest Lapps.*—Dwelling in the coniferous forests that lie to the south of the tundra, the forest Lapps not only fish and keep reindeer but also grow meagre crops.

In winter the Fjeld Lapp lives in a conical tent of reindeer skin with a smoke-hole at the top. Around it graze his reindeer and goats, while dogs warn him of the approach of strangers, who are warmly welcomed, for, like the Eskimo, the Lapp is friendly and hospitable. On the twig-covered floor inside the tent the family, seated upon reindeer skins, encircles a central fire. Scarcity of timber accounts for the absence of chairs and tables.

Lapps wear clothes of reindeer skin trimmed with bright-red

[*Mondiale*

A Herd of Reindeer in Lapland

With their broad hooves the reindeer are scraping away snow to feed on mosses beneath it. Their shadows are long, for even at noon the winter sun is very low in the sky (in mid-winter it does not rise). Note the absence of trees.

cloth. Just as Scottish clans wear different tartans, so various tribes of Lapps are distinguished by the different colours of their caps. Water-proof boots, stuffed with sedge and stitched with reindeer sinews, encase their feet.

Reindeer meat, smoked with juniper, forms an important part of the meal. It is kept on a platform outside of the tent, beyond the reach of the dogs. Barley cakes are sometimes baked, and cheese is made from reindeer milk. Goats, too, provide milk. The goat wanders less and is more easily handled than the reindeer, which must be lassoed before being milked. Hot soup of reindeer blood is given to the dogs.

Like the Eskimo, the Lapp goes fur-trapping in winter. Instead of huskies, however, a reindeer pulls his sledge. The Eskimo must take meat for his dogs, but the reindeer fends for itself by scraping away the snow with its broad hoof and grazing on the mosses beneath.

In spring the return of the sun after its long winter absence

[*Pictorial Press*

The Midnight Sun

After the long arctic winter nights the return of the summer sun is welcomed.
In mid-summer the sun does not set.

is celebrated with joy and gladness. From miles around the
Lapps gather at one church for the Easter Festival. All wedding
and christenings take place on this occasion, while those who
have died during the winter are given a common burial. Summer
and life lie ahead; winter and death have departed.

In summer the Fjeld Lapps follow their reindeer, shepherded
by dogs, up the mountains. To their meals of meat and milk are
now added fish and wild berries. Teachers accompany them and
give lessons in a tent to the younger children. Every week or two
this school moves on a little farther. Among these wanderers
lessons on the plants and animals of their surroundings and on
geography are naturally popular. The Swedish Government has
provided boarding-schools for the older children while their
parents are away in the mountains. They consist of tent-like
huts, where the pupils sit, sleep, and eat, on the floor in preparation for home-life when the family is reunited in the winter.

Civilisation, however, has touched Lapland. Tourists are attracted here by the midnight sun, "lovely as a Lapland night". Model dairies have been established. The iron-ore deposits of Kiruna rank among the richest in the world. Here electric flood-lights enable mining to continue after the early winter sunsets. Unlike Eskimos, however, Lapps are increasing in numbers under the influence of civilisation.

Arctic Russia.—Since the Russian Revolution of 1917 extra-ordinary changes have taken place in the tundra lands of the U.S.S.R. These letters stand for the Union of Soviet Socialist Republics. Rich deposits of coal, iron, oil, nickel, gold, zinc, and apatite for fertilisers have been found. Industrial towns have sprung up, and much of the food needed by their inhabit-ants is grown locally. Swamps have been drained and poor soils made fertile. Frost-resistant varieties of rye, barley, oats, pota-toes, and wheat are grown, and vegetables thrive in the sun-light of the long summer days. Greenhouses yield cucumbers, tomatoes, lettuce, and even grapes, while reindeer, pigs, dairy cattle, and horses are reared.

The women, children, and old people, who formerly lived in tents as they accompanied the reindeer-herders in search of pasture, now remain behind in the permanent houses of *collec-tive* villages. People like the Nentsi, formerly called Samoyeds, which means cannibals, now have schools, hospitals, libraries, chemists, and doctors. No longer need the reindeer-herders journey 500 miles or more in a haphazard search for fodder, since scientists have mapped for them the areas of good pasture. Some herders now work in *collectives*, or groups. Others are employed on State-owned reindeer farms, where food is pro-vided. On their return from hunting and herding, the men exchange at co-operative stores furs and skins for food, clothing, knives, and rifles.

In summer fish caught in the Arctic Ocean are taken to can-neries and fish-oil refineries, and convoys of ice-breakers navi-gate the Great Northern Sea Route which links the Atlantic and Pacific Oceans. Along this waterway have arisen wireless, coal-ing, and weather stations. From the tundra or from the coni-

[*Polar Photos*

A Hot-house in the Arctic

The grapes are growing in the extreme north of Russia. In winter electricity for lighting and heating the hot-house is generated by windmills.

ferous forests to the south new ports export furs, minerals, and timber. At these "mushroom" towns are to be found the world's most northerly hydro-electric power stations, generating power for modern sawmills and metal works.

EXERCISES

1. Keep a record of the very cold belt, and of all the other climatic regions described in the following chapters, by pasting into a scrap-book pictures and descriptions of the climate, natural vegetation, animals, and human life of each region.

2. "All the bread-winners of the tribe are obliged to carry on different occupations at the different seasons of the year." Show how this statement is true of the Eskimo by describing a typical day in his life (*a*) in summer, and (*b*) in winter.

3. Arrange a debate on: "Whether the Eskimos are the Masters of Arctic Nature or her slaves."

CHAPTER XVIII

The Cold Belt

WORLD POSITION AND CLIMATE

IMMEDIATELY to the south of the tundras, the cold belt stretches right across North America and Eurasia. Over western margins the southern limit of the belt is marked approximately by the 60° N. parallel, but over the interiors it sweeps farther southwards and it leaves the continents along their eastern margins at about 50° N. (Fig. 62). In the southern hemisphere, however, continents fail to reach the cold belt.

Summer, although short, is warm, with the mean temperature for the warmest month everywhere rising above 50° F. By contrast, winter is long and cold or very cold (Fig. 64).

Precipitation varies greatly both in amount and in seasonal distribution, as follows:

Position	Annual Amount	Seasonal Distribution
Western margins (Fig. 64A)	Heavy	All seasons.
Eastern margins		
(a) E. Canada (Fig. 64B)	Heavy	All seasons.
(b) E. Siberia (Fig. 64C)	Light	Summer rain, dry winter.
Interior (Fig. 64D)	Light	Summer rain, dry winter.

Throughout the whole belt such precipitation as occurs in winter falls as snow.

NATURAL VEGETATION AND ANIMALS

As the 50° F. summer limit for tree-growth is reached by the warmest month, forests will grow. These are coniferous, for the ordeal of long winter frosts and snows can be faced only by conifers. Thousands of square miles are covered by pine, fir,

165

spruce, hemlock, and larch. To these great softwood forests Russians give the name *taiga*, which means "northern forest". The taiga forms the haunts of the fox, marten, ermine, lynx, bear, beaver, and other furry animals. Some of these creatures migrate northwards from the taiga to the tundra in summer, when food there becomes more plentiful. In Soviet forests hungry wolves sometimes attack human beings in the winter season of food shortage. In fact, Russian airmen are given medals for successful bombing raids on wolves that menace the outskirts of Leningrad.

MAN IN THE COLD BELT

IN these huge coniferous forests man gets a living by selling furs and firs; he either traps animals or fells trees.

The Fur-trapper In North America.—The Canadian fur-trapper is usually either a Red Indian or a French-Canadian descendant of those seventeenth-century French pioneers who, in their search for furs and for a new route to the Far East via the New World, opened up much of Eastern Canada.

Although very healthy, the fur-trapper's life is lonely and often dangerous. Summer is a holiday season, during which he rests or makes himself generally useful at the trading-posts of the Hudson's Bay Company.

In the *fall*, or autumn, he departs by canoe or by sledge for his winter quarters. These may be over a hundred miles away, and consist of a log hut heated by an iron stove and carpeted with twigs. By late autumn the furs are at their best as the animals prepare for the onset of the bitter winter, when trapping yields the finest rewards.

Setting out from his hut, the trapper follows trails along which he sets some forty or fifty steel traps each day. These he visits regularly, re-setting and re-baiting them and dispatching the victims with a blow of his axe. The bodies are piled upon a light sledge and are dragged by his dog-team or by himself back to the log hut, where his squaw skins the animals and dries the furs.

Cloth leggings and moccasins, mittens of animal skin, a fur

Station: KADIAK (Alaska)
TEMPERATURE

Station: QUEBEC (E. Canada)
TEMPERATURE

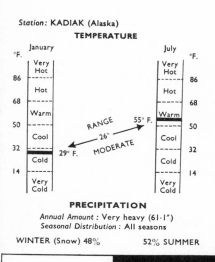
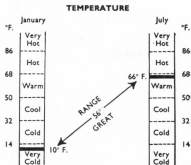

PRECIPITATION
Annual Amount : Very heavy (61·1″)
Seasonal Distribution : All seasons

WINTER (Snow) 48%   52% SUMMER

Depressions at all seasons

A. WESTERN MARGINS

PRECIPITATION
Annual Amount : Heavy (40·7″)
Seasonal Distribution: All seasons

WINTER (Snow) 45%   55% SUMMER

Depressions at all seasons

B. EASTERN MARGINS (N. America)

Station: OKHOTSK (E. Siberia)
TEMPERATURE

Station: CHIPEWYAN (Central Canada)
TEMPERATURE

PRECIPITATION
Annual Amount: Very light (7·5″)
Seasonal Distribution: Summer rain,
     Winter drough

WINTER (Snow) 12% 88% SUMMER

Cold, dry winds On-shore winds
from continental drawn towards
high pressure continental
      low pressure

C. EASTERN MARGINS (Asia)

PRECIPITATION
Annual Amount : Light (13·5″)
Seasonal Distribution : Summer maximum

WINTER (Snow) 33% 67% SUMMER

Cold, dry winds Convectional
from continental showers
high pressure

D. INTERIOR

Fig. 64.

167

[*Pictorial Press*

Stores at a Hudson's Bay Trading Post
At these stores trappers obtain all their requirements. Over 4,000 different kinds of goods are supplied and the mail is also distributed here.

cap, and a leather or fur coat protect the trapper from the cold, and wide snow-shoes support him over the yielding surface of the snow. He carries an axe, gun, and hunting-knife, together with a plentiful meat supply for himself and his dogs, for food in the forest is scarce. At nightfall he either sets up his portable tent or builds a shack of bark, supported upon a framework of strong boughs and open on one side before a blazing fire.

In spring he takes the winter's catch by sledge or canoe to the trading-post. Here, in exchange for his furs, the trapper buys foodstuffs, blankets, a gun and ammunition, a knife, or a gramophone. Consisting of a general stores and a few houses, the trading-post is controlled by a manager, helped by servants. In former times it was strongly fortified against raids by hostile Indians, but nowadays a few cattle and chicken are kept and a wireless set provides the only relief to a peaceful but humdrum

A Russian Fur-trapper   [S.C.R.

This trapper is returning from a hunt. Note his warm clothing, the coniferous forest, and the reindeer searching for food.

existence. From the more isolated posts furs are sometimes carried by aeroplane to the great markets of Montreal and Winnipeg.

Reckless trapping has threatened some animals, e.g. the beaver, with extermination. To safeguard future supplies and to make the fur trade less dependent upon a lucky capture in the trap, fox farms have been established. Here the animals are carefully reared prior to being slaughtered by humane methods.

The Fur-trapper in Asia.—In Russia the fur trade is even more important than in Canada. A bitterly cold winter ensures a huge demand for fur clothing at home, apart from the great export trade.

The industry is well organised. The trappers bring their products to central collective settlements, where modern hospitals, schools, shops, and cinemas have to some extent raised the former appallingly low standard of life.

The Lumberjack.—Besides providing a home for the animals whose furs man prizes so much, these northern coniferous forests are in themselves of great value. The timber is cut down to satisfy man's ever-increasing needs. Some of it is sent southwards to the treeless temperate grasslands, where man uses it to build his house, to feed his fire, and to fence his farm.

Lumbering in North America.—In Eastern Canada forest rangers in summer choose those trees that are to be felled. In autumn a camp is prepared in a clearing. Stores, a bunkhouse, and a cookhouse are built. To offer as little resistance as possible to winter gales, these buildings are squat, but their roofs are steep to cope with heavy snowfalls.

Winter, when the sap is low, is the season for felling. With their branches lopped off, the logs are hauled by horses, by caterpillar tractor, or by lorry over ice-roads to be piled up alongside the frozen streams.

When spring brings the thaw the heaps of logs are broken up and are then tumbled into the rivers to rush pell-mell downstream towards sawmills perhaps 200 miles away. Rapid movement while the rivers are in spate is essential, otherwise logs may be stranded. Sometimes a jam occurs. The key logs must then be prised loose or blown up to clear the passageway, and once movement begins again the men must leap to safety to avoid being crushed to death. Before they are released into the main stream, the logs are hemmed within booms, or chains of logs. The boomed logs, numbering up to 150,000 and sometimes covering the river for a mile or more, form a huge *drive*. Living on a raft, lumbermen follow the drive as it drifts towards the sawmills and pulp and paper mills.

In Eastern Canada, therefore, climate makes transport cheap, for the snow-cover affords an excellent slipway to rivers which, swollen by the spring thaw, give the logs a free ride to the sawmills. For this reason lumbering here is a seasonal occupation. In summer some lumberjacks work in sawmills while others spend their earnings in tasting the joys of city-life after their long exile in the forests. Formerly many would migrate to the prairies to harvest wheat and some still do so.

Lumbering is also important along the north-western coastlands of North America, in British Columbia, Washington, and Oregon. Strictly speaking, this region is not in the cold belt, since it has a temperate maritime climate, with a mild winter and but little snow, save upon mountains. Nevertheless, with coniferous forest instead of deciduous forest as its natural vegetation, it forms part of the taiga belt that crosses North America from the Pacific to the Atlantic Ocean.

Logging railways for the transport of lumber would be unprofitable in Eastern Canada, for throughout the long winter they would be snowbound. In the mild Pacific coastlands, however, they can be operated throughout the year and felling therefore need not be confined to winter. Here the logs are dragged by steel cables or are floated down flumes, sometimes at 20 miles an hour, to light railways which serve the sawmills.

Owing to the enormous number of trees that have been felled and especially to disastrous fires, less than half of the original North American taiga remains. To maintain supplies for the future new trees are planted. In some Canadian schools tree-planting ceremonies are carried out to impress upon the children the need for constant care in the use of Nature's gifts.

Some of the forest has been cleared to make room for mining towns. Eastern Canada supplies over three-quarters of the world's asbestos, nickel, and cobalt, together with much gold and silver. On the borders of Quebec and Labrador huge deposits of iron ore have been discovered, and these will eventually feed steelworks in Canada and the United States. Hydro-electric power stations provide energy for the manufacture of aluminium from imported bauxite.

Railways link the mining towns with ports. In sparsely populated regions railway coaches have been converted into classrooms, and at regular intervals this travelling school visits convenient places along the tracks. In some districts the only link with the outer world is the aeroplane that brings supplies and delivers the mail.

Lumbering in Eurasia.—In Norway, Sweden, and Finland lumbering is carried on in much the same way as in Eastern

[*Swedish Institute*

Timber on a Swedish River

The logs are drifting downstream after the break-up of the ice. A log-boom hems them in. Pulp, paper, and other timber products are exported to many different lands.

Canada. Forest products form a large proportion of the industries and exports of these countries.

The taiga of the U.S.S.R. covers about five per cent. of the total land surface of the world. Here, again, rivers provide the chief means of transport, and it is largely along their valleys that the timber has been cut. As in Canada, mechanical saws fell the trees and tractors haul the logs to the rivers, whose floodwaters in spring float them northwards to arctic sea-ports such as Archangel. Ice-breakers are sometimes used to open up the rivers for transport. Much timber is taken southwards to the treeless steppes.

Based upon hydro-electric power, the largest sawmills in Eurasia have been built at Archangel. Their products are in summer shipped through the Arctic Ocean via the Great Northern Sea Route either to the Atlantic or to the Pacific.

This shipping channel is kept open by ice-breakers, and is fed with timber exports by northward-flowing rivers from Central Siberia. To-day the Siberian forests are being extensively exploited and in the summer steamers can reach places as far inland as Igarka on the Yenisei.

In the Soviet taiga rich deposits of minerals have led to the development of the country, and efforts have been made to feed the increasing population by planting frost-resistant and quick-ripening types of crops.

EXERCISES

1. "For a century and a half during the French rule in Canada the fur trade was at once the mainspring of discovery and development and the curse of settled industry." Explain how the fur trade opened up Canada, but how fur-trapping and permanent settlement could not go together.

2.

Country	Per cent. of Total Area			Acres of Forest per 100 inhabitants
	Forested	Cultivated	Remainder	
Finland	75	6	19	1,470
Sweden	57	9	34	960
Norway	24	4	72	650
Great Britain	5	21	74	10

Represent each country in this table by a column 5 inches high. On each column show the relative importance of forested and cultivated land by shading in different ways the percentage amounts given. On a scale of 1 millimetre to 10 acres make similar diagrams to illustrate the facts given in the last column. Of what kinds of land does the "remainder" percentage consist? Why does Britain differ so much from the other countries?

3. A single edition of the *New York Times* consumes 100 acres of forest. In what ways is man safeguarding timber resources in face of such huge demands upon the forests of the cold belt?

4. Find out what you can about Igarka (page 173) and the all-the-year-round activities of this inland port.

CHAPTER XIX

Cool Temperate Margins

WORLD POSITION AND CLIMATE

THE position of the cool temperate belt is shown in Fig. 65. In northern continents it extends equatorwards from the cold belt almost to the 40° N. parallel. In the southern hemisphere little land extends polewards beyond the 40° S. parallel, so that only Tasmania, the South Island of New Zealand, and the "tail-end" of South America dip into the cool temperate zone.

The mean temperature for the warmest month reaches at least 60° F. in most of this belt. Three sub-divisions can be discerned:

Sub-division	Climate	Alternative Names
1. W. Margins	Temperate maritime	Western cool temperate, Oceanic, British.
2. Interior	Cool temperate continental	Interior cool temperate.
3. E. margins	Laurentian	Eastern cool temperate.

The western and eastern margins will be considered together in this chapter. Cool temperate continental interiors, however, are so unlike either of the marginal regions that they will be described in later chapters, together with the warm temperate continental interiors with which they have much in common.

The temperate maritime climate of western margins (Fig. 66A), with its warm summer, mild winter, and all-seasonal rainfall from on-shore variables and depressions, has been fully described in Chapter XIII. Snow here falls only on mountains and in occasional cold spells.

The climate of the opposite margins is quite different. It is true that summer, as in temperate maritime regions, is warm,

174

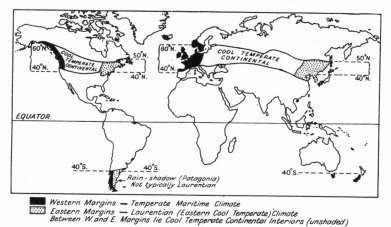

Western Margins — *Temperate Maritime Climate*
Eastern Margins — *Laurentian (Eastern Cool Temperate) Climate*
Between W. and E. Margins lie *Cool Temperate Continental Interiors (unshaded)*

Fig. 65.—Cool Temperate Margins

but in winter these eastern margins freeze up, for they are swept by icy winds from the high-pressure heart of the continents. The range of temperature is therefore great. In their seasonal distribution of precipitation the North American and Asian eastern cool temperate margins differ from each other. Over the Atlantic margins of the north-east of North America depressions yield precipitation throughout the year, and in winter mantle the ground with heavy snowfalls (Fig. 66B). In North-east Asia, however, depressions are less frequent. Here winter is a dry season, with cold north-west winds blowing out from the continental high-pressure centre. Relief rains fall in summer, when these off-shore winds are replaced by inflowing south-east winds, drawn on-shore by the low-pressure system over the interior of Asia (page 102 and Figs. 46 and 66C).

South of the Equator lands in these latitudes are far too narrow for their eastern margins to experience the severe winters of a typical eastern cool temperate climate.

NATURAL VEGETATION AND ANIMALS

WINTERS are everywhere severe enough to check plant-growth. Deciduous forests of oak, elm, poplar, birch, beech, and maple trees clothe the lowlands of temperate maritime regions, where winters are cool and comparatively short. Above them, on the

175

colder mountain-sides, rise "islands" of coniferous forests. In the Pacific margins of North America, however, conifers everywhere take the place of deciduous trees (page 171).

In eastern cool temperate regions, with their long, cold winters, mixed forests of deciduous and coniferous trees are to be found.

In some rain-shadow areas, e.g. Patagonia and the Canterbury Plains of New Zealand, the rainfall is normally too light for tree-growth, and grassland or even semi-desert replaces forest.

Unlike those of dense equatorial forests, trees in these cool temperate woodlands grow sufficiently far apart to allow the free movement of big animals. The moose, elk, deer, bear, and wolf roam here, together with smaller creatures like the squirrel and beaver. These wild animals, however, usually give man's haunts a wide berth.

Man in Cool Temperate Margins

Many great centres of civilisation have arisen in cool temperate margins, where man's activities are numerous and varied.

Farming.—Wherever the forests have been cleared farming is carried on. Mixed farming is the rule. A rotation of crops ensures that the land is not exhausted and is kept in good heart. Apples, cherries, plums, pears, and soft fruits ripen in the warm summers. Fruit orchards are often planted where there is comparative freedom from disastrous night frosts in blossom-time, e.g. near lakes or the sea, or on hill-slopes above misty valley-bottoms.

Vegetables grown near the big towns feed workers in mine and factory. Potatoes are so important in Ireland that in 1845 famine followed an outbreak of blight; they are also grown throughout Western Europe, in Eastern Canada, and in the United States. Sugar is derived from sugar-beet in Europe and from the maple-sugar tree in Eastern Canada.

The chief cereals are wheat, barley, oats, and rye, while flax and, in Eastern Asia, the highly nutritious soya bean are also important.

Root crops and hay provide winter feeding-stuffs for cattle,

Station : MANCHESTER (England)

TEMPERATURE

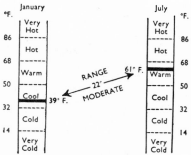

PRECIPITATION

Annual Amount : Moderate (31·7″)
Seasonal Distribution : All seasons

WINTER 46%   54% SUMMER

On-shore variables and depressions at all seasons

A. WESTERN MARGINS
(*Temperate Maritime Climate*)

Station : HALIFAX (E. Canada)

TEMPERATURE

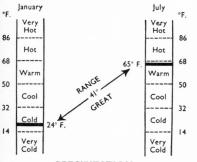

PRECIPITATION

Annual Amount : Heavy (57·3″)
Seasonal Distribution : All seasons

WINTER (Snow) 55%   45% SUMMER

Depressions at all seasons

B. EASTERN MARGINS (N. America)
(*Laurentian Climate*)

Station : VLADIVOSTOK (E. Siberia

TEMPERATURE

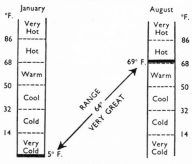

PRECIPITATION

Annual Amount : Light (15″)
Seasonal Distribution : Summer rain,
Winter drought

WINTER (Snow) 17%   83% SUMMER

Cold, dry winds from   On-shore winds drawn
continental high   towards continental
pressure   low pressure

C. EASTERN MARGINS (Asia)

Fig. 66.

177

[*Aerofilms*

Preston, an Industrial Town in Lancashire

Many of the world's greatest industrial regions are in the cool temperate margins of the northern continents.

which are reared for both beef and dairy produce. Pigs, fed upon waste products from the dairy farm, are turned into pork, ham, and bacon.

The sheep-farmer finds the mild winters of the temperate maritime margins more suitable for his flocks than the heavy snows of eastern margins, for sheep, unlike cattle, cannot be penned in sheds throughout a cold winter.

Mining and Industry.—A bracing and invigorating climate, neither too hot nor too cold, has led to the development of great physical and mental powers in the peoples of the cool temperate belt. Wherever minerals occur man has skilfully won them from the earth and upon them has founded busy manufacturing industries.

In North America gold brought settlers to British Columbia, where it is still mined, besides coal, copper, zinc, and lead. Iron from Newfoundland is smelted with coal from Nova Scotia. From New England, rich in water-power, a hive of industry reaches inland to the Great Lakes and to Pennsylvania's famous coalfield, where Pittsburgh repeats the smoke-pall of Britain's Manchester and Birmingham and Germany's Ruhr. By modern methods of smoke abatement, however, the problem of soot in these industrial areas may be largely solved. In fact, compared with its former grimy appearance, Pittsburgh has been transformed into a clean city since the installation of filters into its factory chimneys.

In Western Europe France owns rich iron-ore deposits, while some of the greatest workshops of the world follow a series of coalfields from Great Britain through North France, Belgium, and Germany to Poland. Here the population is sometimes crowded 2,000 to the square mile into grimy cities sprawling beneath sooty skies. Mountain torrents generate electricity to manufacture aluminium in Scandinavia and intricate machinery, clocks, and watches in Switzerland, where skilled craftsmanship overcomes the handicap of a lack of raw materials.

In Asian eastern cool temperate regions great advances in industrial activity have recently been made in the Soviet Pacific regions.

South of the Equator, too, industries are forging ahead. Tasmania has paper mills and zinc, copper, and tin mines, while New Zealand now supplies many of her own needs. Of the southern cool temperate lands only South Chile, handicapped by lofty mountains and isolated from the world's great trade routes, lags behind.

The Harvest of the Sea.—A characteristic occupation of man along the coasts of cool temperate margins is fishing. The four leading fisheries of the world fringe the cool temperate margins of North America and Eurasia, as follows:

Atlantic Fishing-grounds
1. European herring and cod fisheries.
2. Cod fisheries of the Banks of Newfoundland.

Pacific Fishing-grounds
1. Salmon fisheries of the north-west of North America.
2. Japanese and Soviet fisheries.

Compared with these northern fishing-grounds, fisheries in the southern hemisphere are negligible, apart from the antarctic whaling industry.

In the seas that wash cool temperate shores live countless billions of microscopic plants and animals. This *plankton*, i.e. "that which has drifted", provides food for millions of tiny fish, upon which live thousands of larger fish—and so on in a *food-chain*. Plankton thrives in cool and cold water, and it has been called the pasture of the sea, for fish feed upon it just as herbivorous land animals graze on grass.

Surrounding each continent is a submarine ledge, or *continental shelf*, over which the sea is comparatively shallow. Here life-giving sunlight can penetrate to the sea-bed to encourage the growth of fish-food. Naturally, fishing-grounds are particularly important where the continental shelf is very wide, as it is off North-western Europe and Eastern Canada. Fisheries may be classed either as *deep-sea*, e.g. for cod, herring, haddock, and halibut, or as *inshore*, e.g. for lobster, crab, shrimps, and shellfish. Deep-sea fishing is carried on by large steam- or motor-driven drifters and trawlers, which are often owned and maintained by companies. With inshore fishing, however, the individual fisherman, manning his private sailing-smack or motor-boat, comes into his own.

Science now helps the fisherman. By wireless he tells his employers ashore how he is faring with his catch, while the echo-sounder finds shoals of fish for him. Special weather forecasts, too, are issued for the benefit of herring fleets, for wind and cloud play havoc with the catch and storms may cause damage to nets amounting to £1,000 or more. Again, modern fishing craft, equipped with radar sets, can steer safely through fog.

In spite of these scientific aids, the fisherman's life is still dangerous. The fickle moods of these stormy temperate seas formerly led to superstitious beliefs that the ocean was controlled by supernatural beings. In fact, folk-lore still holds a firm grip on many a lonely fishing village, where legends like that of the Flying Dutchman find ready supporters.

Occasionally man has been as rash in his exploitation of the harvests of the sea as in his misuse of the land. Reckless fishing

[*Picture Post*

A Fishing Settlement in the Lofoten Islands, Norway

Fishing is very important in the oceans bordering cool temperate margins. Although the Lofoten Islands are within the Arctic Circle (note the snow on the ground), the sea here is not frozen. Why?

exhausts fisheries. This threat arises particularly in trawling, for the bag-shaped trawl ensnares fish of all kinds and ages, whereas the wider mesh of the curtain-shaped drift net allows the younger ones to escape. Like lumbering, agriculture, and animal-rearing, fishing must be planned if disaster is to be avoided. Where danger of extermination exists control by international agreement is practised, e.g. in whaling and in hunting the fur-seal. Fishermen sometimes neglect fishing-grounds where stocks are failing, to allow them to recover. In some cases young fish are bred in coastal hatcheries and are then let loose into the sea so that future supplies may be guaranteed.

Besides providing him with food, the sea serves man in many other ways. Fishing affords good training in seamanship, and the leading fishing countries possess merchant navies, in some cases out of all proportion to the size of their population.

Thus British, Norwegian, Dutch, Danish, Japanese, and American ships are to be found trading all over the world. The riches of the sea also help the farmer. From waste products of fish are manufactured fertilisers; the Norwegian crushes cod heads into cattle fodder; and the Breton fisher-farmer spreads sea-weed as manure over his fields. Again, the harm done to the health of children by smoke and soot in the overcrowded and sunless cities of Europe's industrial belt is partly offset by the use of cod and halibut oil, so rich in wholesome vitamins.

EXERCISES

1.

	J	F	M	A	M	J	J	A	S	O	N	D
Temp. (°F.)	59	58	55	48	41	35	35	38	43	49	53	56
Prec. (ins.)	0·6	0·3	0·4	0·6	0·4	0·5	0·4	0·6	0·3	0·3	0·4	0·7

The above figures are for Santa Cruz, Patagonia. Draw a climatic diagram and describe the climate. Explain how and why this climate differs from the normal Laurentian climate of eastern cool temperate margins.

2. Compare and contrast the climatic conditions prevailing within similar latitudes on the western and eastern margins of Canada. Account for the differences.

3. During the nineteenth century a flood of immigrants poured into North America from Europe. Why did the great majority of them enter the United States rather than Canada? Of those who went to Canada why were most from Britain or from other North European countries?

4. "All Englishmen are sailors." Show how, because of the position of the British Isles, the sea has always been of supreme importance in the history, commerce, and everyday life of the British people.

5. "An island of forests surrounded by fish." Why is this in some ways an apt description of Newfoundland?

CHAPTER XX

Warm Temperate Margins

WORLD POSITION AND CLIMATE

THE position of the warm temperate belt is shown in Fig. 67. In both hemispheres it extends equatorwards from the cool temperate belt almost to the 30° parallel along western margins and practically to the 25° parallel along the eastern margins of continents.

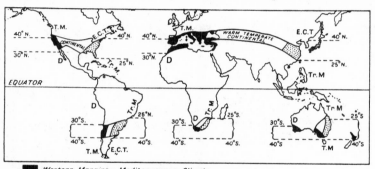

■ *Western Margins — Mediterranean Climate*
▦ *Eastern Margins — China (Eastern Warm Temperate) Climate*

Fig. 67.—Warm Temperate Margins

Between W. and E. margins lie Warm Temperate Continental regions (unshaded). Note how Mediterranean lands link Temperate Maritime (T.M.) to Hot Desert (D) regions, while in N. and S. America Eastern Warm Temperate lands link Eastern Cool Temperate (E.C.T.) to Tropical Maritime (Tr.M.) regions.

Throughout the belt summers are hot, with mean temperatures for the warmest month almost everywhere rising to over 68° F. (Fig. 68A and B, and Fig. 70B and C). As in the cool temperate belt, there are three sub-divisions:

Sub-division	Climate	Alternative Names
1. W. margins	Mediterranean	Western warm temperate.
2. Interior	Warm temperate continental	Interior warm temperate.
3. E. margins	China	Eastern warm temperate.

183

The western and eastern margins form the subject of this chapter, but warm temperate continental interiors are dealt with in the following chapter, together with the cool temperate continental interiors into which they imperceptibly merge.

In seasonal temperatures both western and eastern margins more or less agree. Both experience a hot summer followed by a winter which is cool, and in some parts warm (Fig. 68). In rainfall, however, western are quite unlike eastern warm temperate margins.

Along western margins the Mediterranean climate forms a link between temperate maritime regions and hot deserts (Fig. 67). Consequently it shares some of the rainfall features of each of its two totally different neighbours. We have already learnt that temperate maritime lands throughout the whole year receive rain from on-shore variables and depressions, and that over hot deserts off-shore trades prevail, causing drought at all seasons (pages 112–114). Forming a transitional zone between these two climatic regions, Mediterranean lands, owing to the migration of wind-systems with the swing over the overhead sun, are visited by both on-shore variables and off-shore trades. In winter they resemble temperate maritime margins, with on-shore variables and depressions bringing heavy showers separated by bright sunny intervals. In summer, however, they are like hot deserts, with prolonged drought caused either by off-shore trades or by the dry, high-pressure calms of the Horse Latitudes from which these trade winds originate (page 114 and Fig. 68A).

Situated roughly opposite to Mediterranean lands, eastern warm temperate margins connect eastern cool temperate to tropical maritime regions (Fig. 67). Like Mediterranean regions, these eastern margins receive winter rains from depressions. Unlike them, however, they experience no summer drought. In fact, the reverse is the case, and more rain falls in summer than in winter. The trades, which blow here in summer, are not only on-shore, but are very emphatically so, being drawn ashore, monsoon-like, by the "magnet" of low pressure over the heated interior (Fig. 68B). Heavy relief rains deluge the land as these on-shore winds rise against the high coastal mountains

184

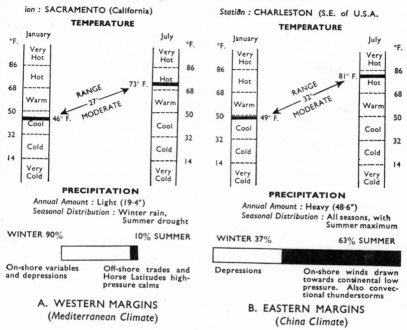

PRECIPITATION
Annual Amount : Light (19·4″)
Seasonal Distribution : Winter rain,
Summer drought

PRECIPITATION
Annual Amount : Heavy (48·6″)
Seasonal Distribution : All seasons, with
Summer maximum

A. WESTERN MARGINS
(Mediterranean Climate)

B. EASTERN MARGINS
(China Climate)

Fig. 68.

that occur everywhere save in South American pampas and in the extreme south-east of the United States. The name " China ", although commonly given to the climate of these eastern warm temperate margins, is not an entirely suitable one, for the northern part of China itself does not experience this climate. It lies not in the warm temperate but in the cool temperate belt, and its winters are dry and cold.

NATURAL VEGETATION AND ANIMALS

WINTERS in warm temperate margins are not cold enough to check plant-growth, and the natural vegetation is therefore evergreen, except, of course, on the colder, mountain slopes.

Mediterranean Vegetation.—The forests of temperate maritime regions gradually thin out equatorwards towards the treeless hot deserts via Mediterranean lands. Thus in natural vegetation, as in climate, the Mediterranean zone provides a transitional half-way stage between its totally different climatic neighbours. Mediterranean regions are not so thickly forested as temperate maritime lands, but neither are they as bare of trees as hot deserts are. Woodlands flourish wherever sufficient rain falls but the trees, like all Mediterranean plants, must resist the summer drought by the methods described on pages 142–143.

Evergreen trees such as the cedar and the cork oak grow on lowlands but on cooler and wetter mountain-slopes are replaced by the deciduous sweet chestnut. Conifers sometimes clothe the higher slopes. In the drier parts are to be found thorny and sweet-smelling plants, e.g. gorse and lavender, while in spring bright flowers form a gay carpet, soon to wither away beneath a glaring sun.

Grass in Mediterranean regions is anything but lush, for in the summer drought it shrivels up and the dry earth cracks.

Eastern Warm Temperate Forests.—In eastern margins the heavy, all-seasonal rains, together with summer heat and an absence of sharp winter frosts, produce evergreen forests of broad-leaved hardwood trees. The mulberry, magnolia, and cedar thrive, often rising from a tangled mass of undergrowth. Eastern Australia is noted for its wealth of tree-ferns and eucalyptus trees, while in South-east Asia bamboo abounds and is indispensable to the Chinese peasant. These jungles have been aptly called "tropical forests in temperate latitudes". Of course, to make room for settlement man has cleared much of the natural vegetation.

In Uruguay and Argentina, however, grassland known as pampas replaces forests which, if they ever existed, disappeared long ago. That trees can grow here has been proved by the planting of wind-breaks around the estancias, or ranches, although during early growth they need protection from the pampero, a strong local wind.

Animal Life.—In both eastern and western warm temperate margins there are numerous animals. Many, however, came originally from other climatic regions, and few can be said to have their natural home here and nowhere else.

MAN IN MEDITERRANEAN REGIONS

CENTURIES ago our western civilisation was born in lands around the Mediterranean Sea. After spreading over most of Europe, it was introduced by European immigrants into many other parts of the world.

The strong rhythm of a hot, dry summer alternating with a cool, wet winter kept early Mediterranean man mentally alert. To farm successfully he had to know exactly when to plough and to sow, as well as how and when to fight drought. In olden days, as at present, farming could not be practised without careful planning. Man was forced to exercise his brain in order to co-operate with Nature, but in return Nature so amply rewarded his efforts that he found himself with sufficient wealth and leisure to enable him to cultivate the art of living to the full a civilised life.

Our debt to these ancient Mediterranean civilisations can never be repaid. To Plato, Aristotle, Socrates, Archimedes, and other Greeks we owe much of our philosophy, mathematics, and architecture. From Roman sources spring our systems of law and government. During the reign of the Roman Emperor Augustus, a period known as the "Augustan Age", literature flourished as it has rarely done since. Later on came the painters, writers, sculptors, and explorers of Italy, Spain, and Portugal— Marco Polo, Dante, Prince Henry the Navigator, Christopher Columbus, Leonardo da Vinci, Vasco da Gama, Raphael, Michelangelo, and others too numerous to mention.

"Lands of wheat and fruit" fittingly describes Mediterranean regions, which agriculturally are among the most productive in the world. Wheat, the chief cereal, finds the climate ideal, with winter rain falling during early growth and summer heat ensuring successful harvests. Abundant summer sunshine and an absence of killing frosts in blossom-time put these regions in the top rank of the world's fruit-growers. The olive, grape, fig, apricot, peach, plum, and nectarine flourish, besides citrus fruits like the orange, lemon, and grapefruit. Many fruits are preserved by canning, or are made into jam or marmalade. Some are, without risk, dried upon trays in the open air.

Many of these Mediterranean fruits were once grown only in the Old World and so, after being gathered in autumn or early winter, appeared in British shops as Christmas luxuries. Later on they were planted in Mediterranean regions in the southern hemisphere. Here they grow during the northern winter and spring, so that South African and Australian grapes and oranges arrive in Britain in summer, well before northern fruits have ripened. Alternating northern and southern supplies therefore enable us to eat fruit throughout the year.

Mediterranean Europe.—In Southern Europe the general standard of living is low. Farms are small and are made to yield as much as possible. The large family lives mainly on what it can grow. The peasant toils from dawn to dusk, although a *siesta*, or rest from work, is taken in the heat of mid-day. This well-established custom is observed in city and country alike throughout Mediterranean Europe. Work on the farm is done by hand rather than by machine, by sickle and scythe, donkey and yoked oxen, rather than by tractor. Machines are expensive and, unlike animals, yield no manure to fertilise the fields.

The peasant cannot afford to waste land. In many areas houses are not scattered haphazardly over the countryside, but are grouped into closely knit villages. These are often perched upon hill-tops, sites which in early days were easily defended from attack and which rose above the malaria-ridden lowlands. Valley-bottoms and plains present a checkerboard pattern of wheatfields, orchards, vegetable gardens and, in North Italy, rice-fields and mulberry trees, upon the leaves of which silk-worms are fed. Irrigation channels lead water along terraced hill-slopes to vineyards and olive groves which at greater heights give way to forests and poor pastures.

As the withered summer grass cannot satisfy dairy cattle, olive oil takes the place of butter. Drinking-water is sometimes scarce in summer, and in areas remote from civilisation may endanger health when exposed in hot weather in unhygienic wells. Wine, made from abundant grape harvests, naturally becomes the national drink. So deservedly famous are wines from Southern Europe that it was some time before "Empire"

[Picture Post

An Orange Grove in Southern Italy

The oranges are taken in carts, drawn by donkeys or horses, to the quayside for export. The Mediterranean climate, with hot, sunny summers and few killing frosts, is ideal for the growing of citrus fruits.

wines from the Mediterranean regions of South Africa and Australia could make any headway in British markets.

The grape-vine survives the summer drought by means of its long tap-root, which seeks out underground water. In fact, too heavy a rainfall would dilute the sugar content of the grapes and so impair the quality of the wine. The plant demands much attention. It is grown upon wooden props and must be sprayed to prevent the spread of disease. Careful weeding is essential, and in South France wind-breaks are needed to protect the vine from the mistral.

The grape-juice is crushed out in a wine-press or sometimes,

189

[*Toni Muir*

A Hill-top Village in Provence, South France

Roofs are almost flat, for in Mediterranean lands rain falls chiefly in winter. Shutters and narrow streets give shade in the hot, dry summer. Hill-top villages were easily defended and are closely knit, so that little valuable land is wasted.

as in Italy, beneath the bare feet of the peasants. The remaining pulp is used as a fertiliser—further evidence of the prudent farming so characteristic of Mediterranean peoples. A useful partner to the grape in the wine trade is the typically Mediterranean cork oak, for its bark provides corks for wine bottles.

Careful labour is also required in olive groves. When the tree is beaten with bamboo poles the olives fall on to sheets spread beneath it. They are then crushed for their oil. Apart

190

from providing a substitute for butter, olive oil is of great value in tinning the sardines that abound in the Mediterranean Sea.

The dried-up summer pastures, although too parched to support dairy cattle, are good enough for beef cattle, while on hillsides sheep and the omnivorous goat provide meat, milk for cheese, wool, and skin for leather. The peasant's plough is drawn by oxen and he piles his goods upon a donkey, mule, or sore-covered horse.

During summer, when growing crops occupy the valley-floors, flocks and herds are driven by their shepherds to graze on hill pastures. After harvest-time they descend to spend winter in the lowlands. This seasonal migration from valley to mountain-side, or *transhumance* (page 160), is practised in the highlands of Eurasia wherever there arises a need for careful management of the available pasture-lands.

To some regions, e.g. the Riviera, the delightful Mediterranean climate brings money, for tourists are attracted there by summer sunshine and by winter warmth.

Non-European Mediterranean Regions.—In the Mediterranean regions of "new" countries farms are usually much larger, and the standard of living is generally higher, than in the Old World coastlands of the Mediterranean Sea.

California, the Atlas region of North-west Africa, Central Chile, the Cape Town area, and the southern tips of Australia (Fig. 67) all produce Mediterranean fruits and wheat. The North Island of New Zealand is an exceptional Mediterranean region, for here there is no summer drought to wither pastures and dairy cattle can therefore be reared. Britain is the best customer for New Zealand's butter and cheese.

Excellent organisation and widespread advertising have put California in the forefront of the world's fruit exporters. In this home of "sunkist" oranges and "sun-maid" raisins "they eat what they can, and can what they can't". Besides important fruit-canning, jam, and marmalade industries, California possesses great oilfields, ship-building yards, and aeroplane factories.

California's sunny climate attracts thousands of tourists, and many Americans have retired there. The population has in recent years increased by leaps and bounds. Particularly in its early stages, before the introduction of arc-lights and indoor studios, the film industry of Hollywood owed much to the clear atmosphere and strong sunlight, so ideal for outdoor photography.

In California and elsewhere in the New World early Spanish settlers introduced their own style of architecture, which obviously shows the influence of the hot, dry summers of Southern Europe. Open, latticed windows are shaded by Venetian blinds or by shutters; light-coloured walls reflect the sun's rays; and within a *patio*, or central courtyard, splashes a cooling fountain.

MAN IN EASTERN WARM TEMPERATE REGIONS

Asian Lands of Rice.—Just as Greece and Rome saw the birth of western civilisation, so China cradled oriental civilisation in the Far East. Chinese scholars could already read and write when Ancient Britons were painting themselves with woad. To Chinese inventors we owe the mariner's compass and gunpowder—which was wisely and harmlessly used for fireworks. When the rest of the world was wallowing in superstitious fears, long before the birth of Christ, the Chinese philosopher Confucius was making known his noble rules of conduct.

The poverty-stricken peasants of China and Japan live mainly upon rice and fish. Rice is easy to grow and is but little affected by disease or insect pests. With most foodstuffs one crop a year is the rule, but two, three, or even more rice harvests are gathered annually in the Far East. A growth of three or four inches a day is not uncommon. A normal farm of two or three acres, little bigger than a large football-pitch, may have to produce food for a family of six or more. No other crop can rival rice in supporting so many people on so little land.

Unlike the wheat and fruit of Mediterranean regions, rice demands heavy summer rains. The menace of soil exhaustion by continual cropping is kept at bay by the application of manure, which keeps the land in good heart and which is highly prized and carefully hoarded. Wading about in the muddy waters of the flooded fields in which the rice is grown, the peasants are frequently crippled with rheumatism. Formerly they grew opium poppies and deadened their pain by smoking

[*Camera Press*
Japanese transplanting Rice from Nurseries to Paddy-fields
Rice is first planted in nurseries, or small flooded fields. When the bright-green shoots appear they are transplanted into larger *paddy-fields*. Paddy is rice in the husk. Rice-straw raincoats protect the workers from heavy rain.

the dried juice of the seed pods. After being threshed with a flail, the rice is cooked with spices added to flavour.

Lowlands, especially river deltas, are best fitted for rice cultivation. The Asian peasant, however, can afford to waste no land, and in mountainous country rice-terraces therefore climb the slopes as vineyards and olive groves do in Mediterranean Europe. These narrow ledges are banked to hold up summer flood-waters, while river mud is carried uphill and is spread over them to provide soil. Hill-slopes also provide ideal sites for tea plantations, as they both cause the heavy rainfall and give the good drainage demanded by the shrubs. Tea is drunk without sugar and, because there are no dairy cows, without milk. Speedy clippers once raced one another to supply Britain with tea from China.

[*Popper*

A Chinese Floating Village

Boatmen peddle food and water, and very many of the inhabitants of these fishing villages seldom go ashore.

Silkworms are fed on mulberry leaves. Much of the raw silk exported from China and Japan goes to France and Italy, whose home supplies cannot maintain the great factories of Lyons and Milan.

Despite her great mineral wealth and the increasing number of her factories, China is not considered an industrial nation. On the other hand, notwithstanding a serious shortage of raw materials, Japan is one of the workshops of the world. The Japanese are more energetic and ambitious than the placid Chinese.

Although in the lowlands and river valleys of China and Japan there is almost "standing-room only", their populations continue to increase at an alarming rate. Here the need for careful farming and for "living-space" is even greater than in Mediterranean Italy. So congested are the river deltas that many Chinese live in boats. To rear cattle or sheep would take up too much valuable space, but pigs and poultry are kept, even on board the boats, since they need little room and eat up scraps

194

of waste food. Roast pig is a great delicacy in China. As there are no pack-animals, goods are carried in baskets hanging from bamboo poles upon the shoulders of human porters, or in creaking wheelbarrows, often rigged with a sail to catch the breeze. Passengers hire a rickshaw although, because of the serious strain upon the puller's heart, this coolie-drawn vehicle is being replaced by the pedicab, a three-wheeled cycle-rickshaw combination.

The Chinese construct their houses and furniture from bamboo, which grows rapidly in the hot, wet summer and which is used to make everything from baskets to boats and from pipes to pagodas. The Japanese house rests, with no foundations, upon the ground. Sliding paper screens form walls that can be removed in hot weather. Outer "rainwalls" of wood keep out the rain. These flimsy houses readily catch fire but are easily and cheaply rebuilt, a great advantage in this land of frequent earthquakes. Modern skyscrapers, however, dominate the centre of the large cities.

Non-Asian Eastern Warm Temperate Regions.—In lands outside of Asia which experience a China climate rice is far less important than maize. Like rice, maize thrives with summer heat and rain and sometimes towers to fifteen or sixteen feet. It is said that on calm nights it may be heard growing.

Although in some areas, e.g. Natal and Mexico, its bright-yellow cobs yield food for man, maize is chiefly used to fatten animals. For this purpose it has few rivals, and its high food-value is thus concentrated in the form of beef, butter, cheese, pork, and lard. In fact, Americans say that maize "is marketed on the hoof" and that it "squeals on its way to market".

Apart from maize-growing, man's activities in the non-Asian eastern warm temperate margins are many and varied, as the following survey shows.

The Deep South of America.—In the *Deep South*, as the south-eastern quadrant of the United States is called, lies the famous Cotton Belt of America. Here the heat-loving cotton bush can be planted without risk of damage by frost, since there are over

[*U.S. Information Service*

Negro Cotton Pickers in the Deep South of U.S.A.

In the Cotton Belt of the South of the United States much of the cotton is picked by descendants of negro slaves.

200 frostless days a year. The chief cotton-growing state is Texas, in the drier west of the belt, for in wetter parts much damage is done by the boll weevil pest.

To liberate Negro slaves in the cotton plantations of the Deep South was one of the aims of the northern *Yankees* when they fought their southern fellow-countrymen in the American Civil War. Cotton is still grown and picked by Negroes but the white pickers now outnumber them. Though slaves no longer, the Negroes are poverty-stricken and are usually in debt. They live in miserably squalid hovels, while the "poor whites" are hardly any better off. Nevertheless, the standard of living is gradually improving, particularly among those employed in the increasing number of factories. The Negroes present a colour problem to which America has so far found no satisfactory solution.

Parts of the American shores of the Gulf of Mexico are too wet for cotton, which here gives way to rice and sugar-cane.

An important occupation along the Atlantic coastal plain is *truck farming*, or the growing of fruit and vegetables on a large scale. The name is derived from the practice of sending the produce by motor-trucks to industrial cities, where it finds a ready sale. Everyone has heard of Virginia tobacco, and "everything is peaches down in Georgia", while sunny Florida rivals Mediterranean California in citrus fruits and in playgrounds for holiday-makers.

The Pampas of Argentina and Uruguay.—From Uruguay and Argentina stream enormous quantities of frozen, chilled, and tinned meat. Millions of cattle and pigs are fattened on maize, while sheep are reared for mutton and, towards the drier interior, for wool. Argentina's flax yields linseed, whose oil is used in the manufacture of paints, varnishes, and linoleum, besides keeping cricket bats in good condition. After its oil has been extracted, linseed is crushed into cattle-cake. Argentina meets half of the world's demand for linseed. Some *estancias*, as farms on the pampas are called, are so large that supervision of farm work must be carried out by aeroplane.

Natal.—Natal grows pineapples, bananas, and sugar-cane on its coastal plain, and tea on its rainy mountain-slopes that face the Indian Ocean. In the sixties of last century Indians were brought here to work in the white man's sugar-cane plantations. Their descendants now add to South Africa's serious problems of poor coloured peoples versus the prosperous white man.

New South Wales.—In South-eastern Australia, as in Natal, a narrow coastal plain is backed by high mountains facing on-shore winds. Here sheep are reared for mutton but not for wool, since the merino sheep that produce wool of fine quality graze farther inland, in the rain-shadow of the mountains. Dairy cows are fed upon maize, and the dairy produce is sold in coastal cities, while much is exported to Britain. Oranges are also important.

Mining and Industry.—Wherever Europeans have settled in eastern warm temperate margins manufacturing industries have arisen. At first local crops and animal products were turned

197

into foodstuffs in fruit canneries, jam factories, butter and cheese creameries, and meat-freezing plants. To these industries there have been added in recent years textile mills, shipyards, engineering workshops, and blast-furnaces. In these new lands there is a tendency for the population to be concentrated around the ports, where the industrial activities are carried on, e.g. at Buenos Aires, the largest city in the southern hemisphere. Manufacturing industries are particularly important if coal is found locally, as at Birmingham in Alabama and in New South Wales, where the most important coalfield south of the Equator has turned Sydney and Newcastle into hives of industry.

EXERCISES

1.

	J	F	M	A	M	J	J	A	S.	O	N	D
Temp. (°F.)	67	67	66	61	57	54	52	52	54	57	60	64
Prec. (ins.)	2·6	3·0	3·1	3·3	4·4	4·8	5·0	4·2	3·6	3·6	3·3	2·9

The above figures are for Auckland, in the "Mediterranean" North Island of New Zealand. In what ways is the climate not typically Mediterranean? Account for the unusual features.

2. Amongst the items in the menu of a Chinese restaurant in London were: (a) rice (with spices), (b) pork and chicken, (c) soya bean cubes, (d) bamboo shoots, (e) tea (without milk). Show how geographical conditions in various parts of China account for each being a possible item in a Chinaman's diet.

3.

	Japan	Italy	U.S.A.
Population per square mile of cultivated land	2,840	1,280	240

(a) Draw three large squares labelled Japan, Italy, and United States respectively. On a scale of 1 dot to 10 people dot the appropriate square to show how the overcrowded margins of much of the Old World, especially the Far East, contrast with the much less densely populated New World. (b) Describe the efforts made in (i) agriculture and (ii) industry and commerce to cope with overpopulation in Japan.

4. How did the Negro Problem arise in the Deep South of North America? Why is it impossible to solve the problem by returning the Negroes to Africa? Try to find out the meaning of the following terms, which are associated largely with the Deep South: Ku Klux Klan, Mason-Dixon Line, Jim Crow Laws, Poor Whites, Share-croppers, and Negro Spiritual.

CHAPTER XXI

Cool and Warm Temperate Continental Regions—1

WORLD POSITION AND CLIMATE

FIG. 69 shows that temperate continental regions are hemmed in between western and eastern temperate margins. They can be sub-divided into: (1) Cool temperate continental, which merge equatorwards into (2) Warm temperate continental lands. In the southern hemisphere, however, land-masses are so narrow wherever they do penetrate the cool temperate belt that the cool temperate continental climate is missing.

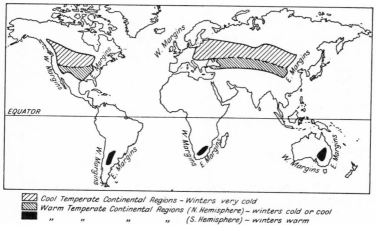

Cool Temperate Continental Regions – Winters very cold
Warm Temperate Continental Regions (N. Hemisphere) – winters cold or cool
  "   "   "   "   (S. Hemisphere) – winters warm

Fig. 69.—Cool and Warm Temperate Continental Regions

Temperature.—If we study Fig. 70A, B, and C, we find that, as befits their position nearer to the Tropics, continental interiors in the warm temperate belt show higher temperatures at all seasons than they do in the cool temperate zone. In North

American and Eurasian interiors, for instance, summers change southwards from warm to hot. Similarly, winters range from very cold to cool or warm (Fig. 70A and B). South of the Equator, however, summers in temperate continental interiors are everywhere hot, while winters are everywhere warm (Fig. 70c). Here the range of temperature is much smaller than in the wider northern continents, where great distance from the sea brings severe winters.

Precipitation.—The annual precipitation is light or moderate, according to how far the area is from the sea. The eastern half of the *Middle West* of America, however, roughly south of the Great Lakes and east of the Mississippi, has a much heavier rainfall than is usual for an interior region.

Everywhere most rain falls in summer and is largely convectional. Wherever, as in parts of the northern continents, winter temperatures drop below freezing-point precipitation falls as snow (Fig. 70A). In fact, an observer often gets a false idea of when most precipitation comes in northern interiors, for the impressive winter snows command far more attention than the much heavier summer rains. We should always remember that:

1. Snow occupies more room than rain, for 1″ of rain equals roughly 10″ of snow.

2. Snow in cold weather evaporates very slowly. Over exposed plains, blizzards tend to blow away the powdery snow-cover, but in sheltered places it lies for months in drifts.

3. Rain from short but heavy summer convectional thunderstorms does not, like snow, outstay its welcome. The water disappears almost as soon as it comes. It is rapidly evaporated, or is carried seawards by swollen rivers, or sinks into the parched ground.

NATURAL VEGETATION AND ANIMALS

ENCOURAGED by a heavier and more evenly distributed rainfall than is normal for continental regions, forests grow in the eastern half of America's Middle West, although man has already cut down most of them   Such forests are exceptional, however, for elsewhere the summer rainfall supports only grass-

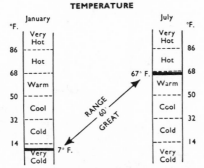

Station : KAZAN (U.S.S.R.)

TEMPERATURE

RANGE 60° GREAT

67° F.

7° F.

PRECIPITATION

Annual Amount : Light (15·4")
Seasonal Distribution : Summer maximum

WINTER (Snow) 27%   73% SUMMER

Cold, dry winds from
continental high
pressure

Convectional
thunderstorms

A. COOL TEMPERATE
CONTINENTAL INTERIOR

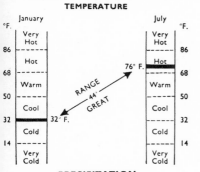

Station : TBILISI (Georgia, U.S.S.R.)

TEMPERATURE

RANGE 44° GREAT

76° F.

32° F.

PRECIPITATION

Annual Amount : Light (19·1")
Seasonal Distribution : Summer maximum

WINTER 34%   66% SUMMER

Cold, dry winds from
continental high
pressure

Convectional
thunderstorms

B. WARM TEMPERATE
CONTINENTAL INTERIOR
(Northern Hemisphere)

Station : C ARLEVILLE (Central Queensland)

TEMPERATURE

83° F.

RANGE 32°
MODERATE

51° F.

PRECIPITATION

Annual Amount : Moderate (21")
Seasonal Distribution : Summer maximum

SUMMER 70%   30% WINTER

Convectional
thunderstorms

Dry (but not cold)
winds from continental
high pressure

C. WARM TEMPERATE
CONTINENTAL INTERIOR
(Southern Hemisphere)

Fig. 70.

F.G. III.—7*

[E.N.A.

A Locust Trap in Argentina

The locusts, before developing wings, are called hoppers and may be trapped by diverting them into a pit. Metal plates overhanging the sides prevent their escape. The insects are burned and the ditch is filled in.

land, as in North American *prairies*, Asian *steppes*, South American *pampas*, Australian *downs*, and the South African *veld*.

The height of the grass depends mainly upon how much rain falls in this summer growing-season. In the drier parts the grassy carpet becomes very threadbare, showing great holes through which bare earth can be seen separating isolated clumps of short, wiry grass. As in tundra lands, spring brings flowers to enrich the scene with bright splashes of colour, but otherwise monotony is relieved only where gallery forests follow rivers. Truly, in these seas of grass "one can see farther and see less than anywhere else on earth".

The fauna is well equipped for life in such wide open spaces. The long legs of the horse and the deer, of the kangaroo and the ostrich, quickly carry them out of reach of their enemies. The foot usually ends not in several toes but in a compact hoof, with its digits fused into one or two only, so ensuring a good take-off with each stride. The rabbit, rat, and prairie dog burrow into the earth for shelter and protection. Other creatures are of the same colours as their surroundings so that, fading into the landscape, they escape detection.

Certain animals, e.g. the horse, cow, and sheep, have been domesticated, and their nomadic owners would indeed be lost without them. A far more difficult task for man than the taming of animals to serve his needs is the conquest of insect pests that make a mockery of his efforts to till the soil. For instance, clouds of locusts, originating from dry areas and blotting out the sun during their flight, settle upon growing crops and consume them to the last stalk. Only too often have they turned a bumper wheat harvest into a dismal failure.

MAN IN CONTINENTAL NORTH AMERICA

BEFORE the coming of the white man the North American Indian roamed the prairies at will. Despising the settled routine of the tiller of the soil, he sought excitement and adventure in hunting bison. In the nineteenth century, however, European immigrants streamed across the continent in their "prairie schooners", or covered wagons. Falling in their thousands before the rifles of these pioneers, the bison became almost extinct, while the Red Indians were eventually defeated and were penned into reservations. Here many of their descendants still live, although they may leave them whenever they please.

Following the trails of these frontiersmen, railways later brought thousands of immigrants from overcrowded and industrialised Europe westwards into the fertile heart of the United States. Germans, Poles, Russians, Irishmen, Swedes, Norwegians, Icelanders, Greeks, and Italians sought in this new land the wealth that the Old World had failed to give them.

Somewhat later on Canada likewise offered glorious opportunities of success. In very early times in Old World Europe towns and villages arose in the most suitable places for settlement, centuries before they were linked up by rail. In other words, European railways followed the people. Across the Canadian prairies, however, people followed the railway. The famous Canadian Pacific Railway was built to join the Atlantic and Pacific Oceans across a practically empty interior. Small townships were then spaced at regular intervals along the railway tracks like beads on a string, for in these monotonous, gently rolling plains no one site holds any particular advantage for settlement over another. Occasionally the hotel, railway station, and bank were built first, and the population would arrive to find their future town-centre already awaiting them.

In fact, by linking together a hotch-potch of different nationalities railways did much to prevent the rise in the New World of separate Germanies, Italies, Norways, Russias, etc. They helped to throw the various foreign peoples into a huge "melting-pot" from which were to emerge patriotic Americans and Canadians.

The Wheat Farmer of the Prairies.—Having ousted the bison and Red Indians, the pioneer turned the prairies into huge cattle and sheep ranches. Later on wheat was grown on a large scale, particularly during the World War of 1914–18, when high prices brought prosperity to the farmer.

Unfortunately, ploughing and overgrazing, i.e. the rearing of too many animals upon pasture-lands, stripped off the soil-binding cover of grass. Bare earth lay exposed to sun, rain, and wind, and the soil, a product of centuries of rock-weathering, was washed or blown away. Moreover, prairie fires had enriched the land, for the ashes provided an excellent fertiliser. The farmer, however, checked these fires, and by growing wheat year after year drained the land of its fertility. "Go west, young man", a popular cry during the opening up of the prairies in the nineteenth century, became an out-of-date and mocking slogan in these man-made deserts.

With thousands of square miles of once fertile land ruined, man has at last learned his lesson, and the mixed farming that is typical of Europe is spreading over much of the prairies. Moreover, as the population increases this kind of farming becomes essential, for there is a growing demand for locally produced butter, milk, eggs, fruit, and meat.

The three Canadian prairie provinces of Manitoba, Saskatchewan, and Alberta together form one of the greatest granaries in the world. Across the prairies rainfall decreases from east to west and in Manitoba, the most easterly and most densely populated of the prairie provinces, it is heavy enough for the mixed farming that is characteristic of well-peopled areas. Saskatchewan, the central province, is the leading

[*International News*

Victims of Drought and Soil Erosion
Soil erosion has devastated much of the once fertile prairies. These Nevada cattle have starved to death.

wheat producer. To the west lies Alberta, much of which is too dry for wheat cultivation without the aid of *irrigation* or *dry farming*.

Irrigation means the moistening of soil with water from ditches. These are usually fed from a reservoir held up behind a dam built across a river, although, as we shall see later, there are other methods of irrigating crops. Dry farming is practised by allowing land to lie fallow for a year or more while the rain-water which sinks into the soil is carefully conserved by harrowing the surface and by keeping down the weeds. In the following year a crop is then grown with the accumulated rainfall of the two or more years. As different fields are fallow each year, farms where this method is used are bound to be large.

In the cooler north of Alberta, where the growing-season is very short, quick-ripening types of wheat make the most of the long summer daylight which compensates for lack of heat.

Taking the prairies as a whole, wheat flourishes because the climate admirably suits its growth. After harvest-time ploughing and sowing cannot be completed before winter frosts harden the ground. Winter is therefore a season of rest, when the severe

205

[*U.S. Information Service*

Strip Cropping on the Prairies
To check soil erosion, to preserve fertility and to combat drought, Canadian and American farmers often plough only half their land, especially in the drier regions. Alternate strips lie fallow for a year.

cold aids the farmer by killing pests and breaking up the soil. If he has earned good money from a bumper harvest, the farmer can afford to take his family for a holiday in the city during the winter. Seed-time is delayed until spring, when melting snows moisten the fields to give growth a good start. Responding to rain in early summer, the ear rapidly heads out. Later on, convectional showers and a glaring sun urge on growth until the ripened grain awaits harvest.

As he can offer little work during the long, cold winter, the farmer cannot afford to employ many permanent farm-hands. At harvest-time machines must therefore be substituted for human hands. Moreover, prairie farms are large. They are not divided into the checkerboard pattern of small, hedge-bound fields so typical of a British farm, and without huge machines the harvests could never be reaped. Great combine-harvesters cut, thresh, and bag the wheat. Manned for this busy season by a temporary labour supply, they are driven from farm to farm. They are usually owned by companies.

A spider's web of railway lines covers prairie wheatlands, for the crop pays only if it is grown within twenty miles of a

[E.N.A.

Rounding up Cattle on an Indian Reservation

These herds are owned by Indian cowboys living in a reservation in Nevada, U.S.A. The Government has encouraged tribal enterprises, including cattle-rearing, and the Indians repay Government loans with cattle.

railway station. Here the grain is stored in an elevator, nick-named the "prairie cathedral", from which it begins its long journey to overseas markets.

Naturally this great spring wheat belt crosses the 49th parallel into the United States, where Minneapolis possesses the largest grain exchange in the world and grinds over 40,000 sacks of flour a day. Buffalo and Kansas City, however, have now out-stripped Minneapolis to become the world's leading milling centres. America grows far more wheat than Canada but can spare less for export, since there are about thirteen Americans to feed to every Canadian.

The Rancher of the High Plains.—The high western prairies that flank the Rocky Mountains are called the *High Plains*. Here a dry climate, made even drier by warm Chinook winds, makes beef cattle ranching a more profitable occupation than wheat-growing.

Cattle once ranged the open countryside at will. Some still do so, but fences are now a feature of the landscape. In many parts the tough natural pastures have been replaced by alfalfa, a nutritious plant whose deep roots find water even in dry lands. The herds are rounded up by cowboys on horseback or in motor cars. Grazing as they go, the future meat supplies are slowly driven to the nearest railhead. From here they travel to the richer lowlands farther east for a final fattening before they meet their doom in the slaughter-houses of great meat-packing cities. Although Americans eat more meat than most peoples, there is a big surplus for export.

The Farmer of the Middle West.—East of the High Plains and south of the Great Lakes, and drained by the Mississippi and its tributaries, lies the world's biggest and richest continuous area of farmland, commonly called the Middle West. Southwards it fades into the Deep South (page 195).

As has already been mentioned, much of the region, especially towards the east, receives a heavier and more evenly distributed precipitation than is usual in continental interiors. In winter deep snows mantle the northern half.

Summer temperatures are high and convectional storms are severe. "The sun beat down upon us with a sultry, penetrating heat almost insupportable. . . . At last, towards evening, the black heads of thunder-clouds rose fast above the horizon, and deep mutterings of distant thunder began to roll hoarsely over the prairie. . . . The thunder here is not like the tame thunder of the Atlantic coast. Bursting with a terrific crash directly over our heads, it roared over the boundless waste of prairie, seeming to roll around the whole circle of the firmament with a peculiar and awful reverberation. The lightning flashed all night, playing with its livid glare upon the neighbouring trees, revealing the vast expanse of the plain, and then leaving us shut in as if by a palpable wall of darkness" (Francis Parkman).

Temperatures rise steadily southwards. The crops grown range accordingly from spring wheat, hay, and apples in the north, southwards through corn (as Americans call maize) and winter wheat, until the Middle West merges into the Deep South, with its Cotton Belt and Gulf Coast zone of rice, sugar-cane, and sub-tropical fruits (see Fig. 71).

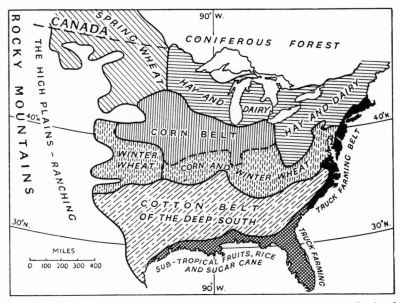

Fig. 71.—The Agricultural Belts of the Middle West and Atlantic Coastlands of North America

Mining and Industry in Continental North America.—In the interior of the United States some of the world's largest deposits of copper, coal, iron, and oil have given rise to densely populated islands of busy mining towns and teeming industrial cities which interrupt the sea of crops. Alberta, too, has prosperous coalfields, while recent discoveries there of astoundingly rich oilfields will put Canada in the forefront of the world's oil-producers.

EXERCISES

1. Contrast the life of a wheat farmer in Saskatchewan with that of a farmer in East Anglia.
2. Arrange a debate on whether man's interference with Nature has been beneficial or otherwise.
3. "Railways are to the United States what ships are to the British Commonwealth and caravans to the Near East." "The Canadian Pacific Railway is more than a mere railway. It is one of the greatest colonising agents in history and an Empire builder with few equals." What do you think is meant by each of these quotations?

CHAPTER XXII

Cool and Warm Temperate Continental Regions—2

MAN IN CONTINENTAL EURASIA

The Russian Farmer.—Work for the wheat-grower of the Soviet steppes follows the same seasonal rhythm as for the North American prairie farmer. Here again winter snows bring farm work to a halt. Spring ushers in seed-time, although in the south-west of the Ukraine winter wheat is planted. In a soil moistened by melting snows, the wheat rapidly matures beneath convectional showers and a hot sun.

In their management, however, Soviet farms are worlds apart from the privately owned farms of Canada and the United States. Farms in Soviet Russia fall into two groups:

(a) *State Farms.*—The Government owns these huge concerns and the labourers receive a weekly wage. On each one a team of scientists experiments with new types of seeds and new methods of cultivation, while mechanics learn how to drive, repair, and maintain tractors and machinery. The total area of the 4,000 State farms equals that of England and Wales.

(b) *Collective Farms.*—Owned by a community or by a collection of communities, these farms cover an area more than three times that of the British Isles. They number almost a quarter of a million. Being the common property of the people, the collective is controlled by a committee, or council, which is elected by the workers. The word *Soviet*, in fact, means Council. Combine-harvesters and tractors are obtained from State depôts, and scientists show the workers how to make the best use of their land.

Each man is given the work for which he is best fitted. The amount of labour which each day can reasonably be put

[S.C.R.

Harvest Time on a State Farm near Rostov in the Ukraine
The continental summer is hot—note the sunshade on the harvester. The electric lights enable work to be continued at night. Why is this necessary?

into the worker's task, including office work and the maintenance of machinery, is called a *labour day*.

When the crops have been harvested the Government first claims its share of them to pay for the use of tractors, for insurance, and for advice given by its experts. Part of the harvest is then stored to provide seed for the next year, and part is sold to meet expenses such as State taxes and the upkeep of community clubs, barns, cinemas, and parks. The remaining crops are shared out among the workers strictly in proportion to the number of labour days standing to each man's credit. Sometimes a crop is sold, in which case wages are paid in cash.

For his personal use the peasant is allowed to own a house, a vegetable garden, and a few animals. At the local market he may sell produce from his own plot of ground as well as his share of crops from the collective. Obviously, the harder he works the more he earns, but it is said that some peasants tend to neglect their duties on the collective farms in order to make money by producing more from their personal allotment.

211

To discourage such private enterprise, the Government puts a heavy tax on all income earned in this way. On some collective farms the peasant is persuaded to work harder by a scheme of piece-work, i.e. he earns money for each task successfully completed, instead of waiting for a share in the harvests on a basis of labour days.

The Nomadic Herdsman.—Among the scores of different peoples who make up the Union of Soviet Socialist Republics are the Khirgiz, Kazaks, Uzbeks, and Turkmens, of Central and South-western Asia. These nomadic peoples wander in a never-ending search of pasture for the cattle, sheep, goats, camels, and horses that serve their everyday needs.

A nomadic tribe consists of a group of separate families, each of which is ruled by its elderly *patriarch,* whose word is law. Pastures are held as common property by the whole tribe, although each family possesses its own animals. This age-old way of life is described in the Old Testament.

The Khirgiz and Kazak nomads in winter take shelter in valleys at the foot of the snow-capped mountains that form the southern boundary of the Soviet Union. Here grass and mountain streams nourish their flocks and herds. In summer the Khirgiz drive their animals southwards into the mountains in search of pasture, whereas the Kazaks trek northwards into the steppes, a change of feeding-grounds which sometimes means a journey of 500 or 600 miles.

Constantly on the move, these nomads find heavy personal property a handicap. The men count their riches in the number of animals they possess, and the women in their elaborate, gem-studded head-dresses. Also highly prized are leather top-boots, carpets, and interior decorations for the *yurt,* or tent. The men wear tall, shaggy hats of sheepskin and wrap themselves in a cotton gown, well-padded in winter to keep out the cold.

The yurt is made of woollen felt stretched over a framework of willow rods, and is bound together with ropes. A hole in the top serves as both window and chimney, letting sunlight in and smoke out. By day warm sunshine also floods in through the door, which always faces south. Dried dung, the only fuel,

212

An Encampment of Mongolian Nomads

[E.N.A.

The wooden door of the foremost yurt is a luxury. The securely bound, dust-proof felt covers a lattice framework and withstands storms. From the word *ordu*, a group of yurts, is derived the word *horde*, first applied to Mongolian armies.

feeds the fire. At night, with feet outstretched towards a central bowl of hot ashes, the whole household sleeps beneath an enormous family quilt.

When the day's work is done the family unites for the evening meal. The men have been riding far afield, driving to the pastures their flocks and herds. The nomadic steppe-dweller must be an expert horseman. No sooner do children leave the cradle than they are in the saddle, and the Kazak, or Cossack as we commonly call him, is world-renowned for his horsemanship. While the men have been away the women have been cooking meals and milking goats and mares. To them also falls the task of dismantling the yurt and setting it up again whenever a move is made to new pastures. Now, as night falls and the animals are safely penned within enclosures, all settle down to a feast which soon develops into an endurance test for appetites. The first course consists of soup made from China tea brewed in hot water with flour, salt, milk, and butter. This dish is followed by *koumiss*, a drink of mare's milk fermented in a goatskin bag. Next on the menu comes mutton stew and

213

roast fat from sheep-tails. The nomad does full justice to this banquet, for during the day he has merely munched the seeds of pine-cones and of water-melons, drunk tea-soup, and taken a little snuff.

To waste food is considered a crime, for in these lands of uncertain rainfall one can never tell when grass may wither and flocks die. In times of want nomadic herdsmen formerly raided the fields of neighbouring crop-cultivators, but a strong Government now frowns on such breaches of the peace. Centuries ago hordes of these restless Asian wanderers invaded Central Europe. In Hungary their Magyar descendants still rear animals and perform daring feats of horsemanship on the flat, grassy *puztas*.

With Government assistance, many nomads have given up their roaming life and have settled down to grow crops. In dry areas irrigation schemes and drought-resistant seeds ensure successful harvests, and Kazakhstan now rivals the Ukraine as the chief granary of the Soviet Union. A nomadic way of life tends to increase soil erosion, for under constant grazing the grass dies and the soil, trampled into dust by countless hooves, is blown away. The danger has now been lessened by the introduction of mixed farming and by the planting of trees, which not only bind together the soil but also abate the force of destructive winds. Wide gallery forests are being planted along 5,000 miles of riverside, and a cantata has even been composed in praise of Soviet afforestation schemes.

Science now aids the herdsman. Over huge collective ranges improved breeds of cattle graze on well-kept pastures. Pigs and cattle are fattened on maize before being sent by train to meat-packing cities like Novosibirsk, the "Chicago of the Soviet Union". In all but the driest areas nomadism is becoming unnecessary. Nevertheless, for those who still prefer to roam the Government has established collective farms along the usual nomad routes. These serve as halting-stages to help the wanderers on their way.

Mining and Industry in Continental Eurasia.—Mineral wealth in Eurasian steppes and semi-deserts has not been neglected.

As in the interior of the United States, thriving industries are based upon rich deposits of coal, iron, copper, and especially oil. Brand-new towns have sprung up like mushrooms. Chemicals made in the cities fertilise the farms, which in turn yield cotton and other raw materials for city mills. Iron and manganese are made into steel products such as agricultural machinery. By helping the farmers in these ways the miners and factory workers indirectly help themselves, for the produce of the fields feeds the cities. Here, as in all great industrial areas, the town depends upon the country, and the country upon the town. Many a former nomad is now a mechanic, although he may still occupy his traditional yurt, even in the town, where it provides a strange contrast to modern buildings. Roads and railways now cross steppe-lands which until recently were linked with the outer world only by camel caravan.

These new developments in mining, industry, and scientific farming have brought about certain improvements in the standard of living. In some parts the modern steppe-dweller can read and write; his illnesses are treated in hospitals, and his children attend schools. In the new towns theatres, cinemas, libraries, and museums serve to entertain or to instruct him. His old customs are not forgotten, however, for folk-dances, national costumes, and native arts are actively encouraged in festivals and exhibitions. In fact, instead of being vigorously controlled by his surroundings, man in the steppes is learning how to master and to improve them.

The Lamas of Tibet.—Beyond the mountainous southern boundary of the U.S.S.R. stretch the high plateaus of Tibet and its neighbours. In these barren lands, cut off from the sea both by distance and by the loftiest mountain barriers in the world, precipitation is very light. So bleak are some parts that no vegetation will grow, although in sheltered valleys millets, maize, and wheat flourish with the aid of irrigation.

The superstitious Tibetan peasant works like a slave for his overlord, and is totally out of touch with the civilised world beyond his mountainous horizon. He depends largely upon his yak. Sure-footed and covered with long hair, this creature is

[E.N.A.

Ploughing with Yaks in Tibet

Tibetans eat yak flesh, butter and milk. The hide is made into ropes and tents, and the tail is exported to India as a fly-whisk. The yak is also a pack-animal.

ideally fitted for life among bleak, mountainous crags and precipices. It gives milk and meat, and as a beast of burden can carry up to three hundredweights.

The most densely populated part of Tibet is the wetter south-east, to which monsoon rains from India can penetrate. Here lies the capital, Lhasa, where the Dalai Lama, ruler of the country, normally lives in state. Many of the men are *lamas*, or monks, who worship devils and perform weird rites. Prayers are written on wheels or on banners, and are believed to be safely delivered to the spirits whenever the wheels are turned or the banners flutter in the breeze.

From the Gobi Desert and Tibet these mountainous wastes stretch westwards and south-westwards, through Afghanistan and Iran, to merge with the hot deserts of Iraq, Arabia, and the great African Sahara. A huge continuous desert thus reaches all the way from the Atlantic Ocean into the heart of Asia, and is broken only by the long, narrow, well-populated oases created by Egypt's Nile and Iraq's Tigris-Euphrates.

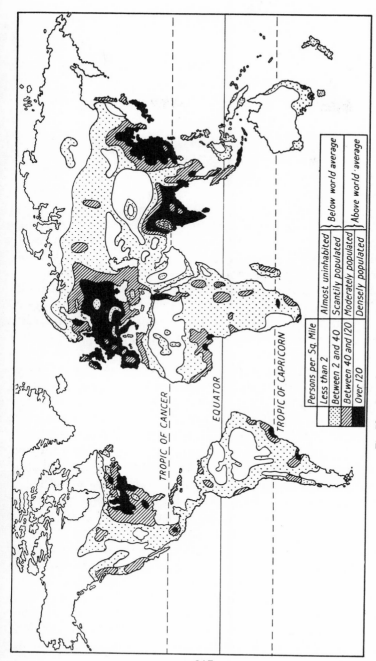

Persons per Sq. Mile

Less than 2	Almost uninhabited } Below world average
Between 2 and 40	Scantily populated }
Between 40 and 120	Moderately populated } Above world-average
Over 120	Densely populated }

Fig. 72.—*Distribution of Population over the World*

TROPIC OF CANCER

EQUATOR

TROPIC OF CAPRICORN

This desert belt sets up an unpopulated barrier between the swarming millions of industrial Europe and the even more crowded agricultural lands of South-east Asia (Fig. 72). It is this desolate barrier which effectively separates the white from the black, brown, and yellow races.

WARM TEMPERATE CONTINENTAL LANDS SOUTH OF THE EQUATOR

IN the southern hemisphere the Murray-Darling Basin of Australia and the interior of the pampas of Argentina repeat the pattern of human activities of northern temperate continental lands. The same kind of highly mechanised farming solves the same problem of shortage of labour in the same sort of huge, hedgeless wheat-fields. There is the same haulage of wheat by road and rail to ports for export. As winter is definitely warm, however, there is no need for the farmer to wait until spring before he sows his seed. In dry parts there is the same need for irrigation, dry farming, and drought-resistant seeds. There are the same dust-storms, and the same problems of soil erosion and insect pests.

There is the same raising of beef cattle on tough native pastures or on deep-rooted alfalfa, and the same long trek to wetter regions for a fattening on maize before slaughter, refrigeration, and export. Again, Australians and Argentines, like Americans, are fond of meat. There are the same millions of sheep, grazing on native grasses or on pastures improved by man. Southern sheep-farms feed the mills of Huddersfield and Bradford with wool and their workers with frozen mutton.

In South Africa, however, the veld differs from Australian downs and Argentine pampas in that wheat is unimportant there, while the native-owned cattle, being of poor quality, at present provide no meat for export. Nevertheless, plans have been made to produce large quantities of excellent beef from Bechuanaland and other parts of Africa (see page 258).

Finally, wherever mineral wealth has been discovered, e.g. gold and diamonds in South Africa and silver, lead, and zinc in Australia, there are the same islands of urban population interrupting the sea of farmland. Surrounding these towns,

An Australian Sheep Station   [*Mondiale*

These merino sheep are being brought in from the New South Wales back-country for shearing.

and supplying their needs, lies the same zone of mixed farm-ing, with orchards and market gardens.

EXERCISES

1.

	J	F	M	A	M	J	J	A	S	O	N	D
Temp. (°F.)	69	70	68	63	59	55	55	56	57	61	64	67
Prec. (ins.)	0·7	0·6	0·9	1·8	3·9	4·4	3·5	3·3	2·2	1·6	1·1	0·8

These figures are for *one* of: Sydney (34° S.), Cape Town (34° S.), Astrakhan (46° N.). (*a*) Describe the climate. (*b*) State, giving reasons, to which place the figures refer. (*c*) How would you expect the figures for the other two places to differ from those given?

2. Russia is sometimes described as the coldest country in Europe, Italy the sunniest, Spain the driest, and Norway the wettest. Explain why these descriptions are appropriate.

3. The U.S.S.R. and the U.S.A. are often said to be alike. In what ways is this statement both true and false?

CHAPTER XXIII

The Hot Belt—1. Tropical Monsoon Lands

THE HOT BELT AS A WHOLE

THE hot belt extends both northwards and southwards from the Equator until it merges into the warm temperate belt at about 30° N. and S. along western and at about 25° N. and S. along eastern margins (page 183).

Taken as a whole, the hot belt may be said to consist of "Lands of Palms". Among the most important rank the date, oil, sago, and coconut palms, all of which demand high temperatures throughout the year.

Temperature.—Throughout much of the belt temperatures are always high, and the annual range is often far too small to make worth while any attempt to distinguish seasons on a basis of summer versus winter. With a noonday sun always high in the sky, and in most parts overhead twice a year, such persistent heat is to be expected.

In most cases mean monthly temperatures exceed 68° F. at all seasons, so that even "winter" is hot, while "summer" is sometimes very hot, with temperatures of over 86° F. The most noticeable seasonal changes in temperature occur in cloudless hot deserts, where the annual range is moderate and occasionally is great (Fig. 76). On the other hand, in tropical maritime and equatorial climates the seasonal rhythm is particularly weak and the annual range is negligible (Figs. 75 and 80). Important exceptions to these temperature conditions occur:

(*a*) Along cold-water coasts, where cold ocean currents cause cooler weather.

(*b*) In hot deserts and some monsoon regions, where winters are merely warm, with temperatures falling below 68° F.

(*c*) On highlands, which naturally give cooling relief from the sultry heat of the lowlands.

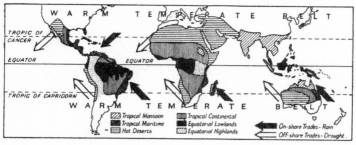

Fig. 73.—The Regions of the Hot Belt

Precipitation.—In the hot belt precipitation falls either as relief rains from on-shore monsoon and trade winds, or from convectional thunderstorms caused by intense heat in the low-pressure Doldrums belt or elsewhere. Depressions do not reach these latitudes. Variations in the seasonal distribution and type of rainfall enable us to distinguish five separate types of hot climates:

Climate	Rainfall		
	Season(s)	*Type*	*Cause*
1. Tropical Monsoon	Summer	Relief	On-shore monsoons.
2. Tropical Maritime	All	Relief	On-shore trades.
3 Hot Desert	None	—	Off-shore trades.
4. Tropical Continental	Summer	Convectional	Doldrums in parts next to equatorial regions. Summer heat elsewhere.
5. Equatorial	All	Convectional	Doldrums.

The world distribution of these five climates is shown in Fig. 73.

TROPICAL MONSOON LANDS—WORLD POSITION AND CLIMATE

THE tropical monsoon climate is experienced in South-east Asia, along the northern fringe of Australia, and in Abyssinia (Fig. 73). The Guinea Coast of West Africa, although it receives heavy rains from on-shore south-westerly monsoon winds, does not experience the long, dry season of a typical monsoon climate (page 102).

221

Station : DARWIN (N. Australia)

TEMPERATURE

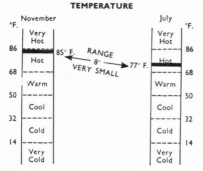

Note that the hottest month is not January, but November, just before the monsoon bursts and skies become cloudy-covered

PRECIPITATION

Annual Amount : Very heavy (61·7″)
Seasonal Distribution : Summer rain,
Winter drought

SUMMER 94%   6% WINTER

On-shore winds   Dry, off-shore winds
drawn towards   (normal trades) from
low pressure over   high-pressure Horse
interior   Latitudes

Fig. 74.

In summer very low pressure prevails over the heated land, drawing ashore monsoon winds which yield torrential relief rains (page 97 and Fig. 46). In causing discomfort these winds sometimes rival the moist heat. Indeed, it has been said that monsoon gales, rushing fiercely through the Palghat Gap in South India, are partly responsible for the very high murder-rate in this part of the country, so violent are the quarrels resulting from the frayed nerves of the inhabitants. Many Europeans who live in Pakistan and India seek relief in cooler Himalayan hill-stations from the moist heat of the lowlands.

The rainfall is heaviest wherever highlands face these saturated on-shore winds. Thus in June, 1950, there was a rainfall of 43 inches in under 48 hours in parts of the Himalayas in Western Bengal. In mountainous Assam lies Cherrapunji, the wettest place on earth. It receives over 450 inches of rain a year—and well over 90 per cent of it falls in the summer half-year! In 1861 Cherrapunji was deluged with 905 inches, of which 366 fell in July alone. By contrast, Lahore, situated well inland on lowlands, receives only 18 inches.

Such torrents falling from cloud-covered skies are bound to cool the air in the rainy season. Consequently temperatures are highest early in summer, before the monsoon bursts, rather than around midsummer.

In winter the normal trade winds of these latitudes reassert their sway and, being off-shore, cause drought. Three seasons can therefore be recognised: (i) a warm, dry winter, (ii) a very

[*E.N.A.*

A Street Scene in Bombay

In summer the umbrella affords protection against both hot sunshine and heavy monsoon rains. Find other evidence of heat in this picture.

hot, dry season that becomes oppressively "sticky" just before the monsoon bursts, and (iii) the hot season of the rains.

The directions of these reversing summer and winter monsoon winds are given on pages 98 and 100 and in Fig. 46A and B.

NATURAL VEGETATION AND ANIMALS

DENSE forests of many kinds of evergreen hardwoods clothe the wettest parts, where the soaked soil retains enough water to maintain trees through the dry winter. With rope-like creepers clinging around the trees or hanging from their boughs, these monsoon rain forests resemble those of equatorial regions.

Where, because of a lighter rainfall, the soil is less saturated, leaves must be shed to check transpiration during the dry season. Here, therefore, evergreen forest gives way to deciduous trees scattered over grassland, which is the normal natural vegetation in lands of summer rain and winter drought. The best known of the deciduous trees is the teak, which flourishes

223

in India and Burma. In the driest parts vegetation is reduced to thorny scrub.

When, after the prolonged dry season, the monsoon bursts "so instantaneous is the response of Nature to the influence of returning moisture, that in a single day, and almost between sunset and dawn, the green hue of reviving vegetation begins to tint the saturated ground" (Sir J. E. Tennant).

Animal life abounds. In the tree-tops monkeys, cockatoos, and brightly coloured parrots chatter and squawk as below them the elephant crashes through the jungle, sometimes trampling under foot man's crops and huts. Sometimes man's life itself is endangered, for leopards and man-eating tigers slink through the undergrowth, while the death-rate from snake-bite is high. Insects and blood-sucking leeches are as numerous as they are irritating.

MAN IN TROPICAL MONSOON LANDS

Tropical Monsoon Asia.—The arrival of monsoon rains is a matter of life or death for the Indian *ryot*, or peasant. The failure of the monsoon once spelt certain starvation for thousands, but nowadays railways rush food from areas of plenty to those of want. Nevertheless, the threat of famine remains, despite modern dams and reservoirs for storing water to moisten a thirsty land in times of crisis.

Under summer heat and rain the soil yields such quick returns that tropical monsoon South-east Asia supports over 500 million people, just as temperate monsoon Asia does in China and Japan. To feed these hordes of human beings the land is divided into tiny plots, necessitating the use of hand tools such as the spade, hoe, and animal-drawn plough. Manure is greatly valued, for it helps the ryot in his efforts to extract a great amount of food from a small amount of soil.

In India scraggy cattle and goats provide the Hindu with milk but not with meat, for his religion forbids him to kill animals, although nothing prevents him from ill-treating them. The donkey, ox, or water-buffalo pulls his plough and his cart, or works crude machines for raising water from river and well to fill irrigation ditches.

[E.N.A.

Ploughing a Rice-field with Buffalo in Ceylon
Heat and heavy rains make possible the growth of several crops of rice a year.

The tea-pots of Britain, Australia, and other tea-loving nations are filled from the plantations of Assam and Ceylon. Although it demands a heavy rainfall, the tea-shrub dislikes stagnant water around its roots. It is therefore planted on rainy but well-drained hillsides. To pluck the crop, or *flush*, of tiny leaves is a delicate task calling for the nimble fingers of women and children, although a tea-picking machine has recently been invented and is making headway in some areas. Heat and rain combine to produce a flush about every ten days, and some plantations yield twenty flushes a year.

India's ever-increasing needs at home have led to a decline in her exports of certain crops. For instance, after feeding her own peoples she has no rice left for overseas markets, and British rice puddings are made from the surplus that Burma can offer for sale. Bengal, which is shared between India and Pakistan, produces practically all the world's jute, while from both India and Burma comes teak, an invaluable timber where resistance to rot is called for, e.g. in ship-building and for lock-gates. The teak logs are dragged from forest to riverside by elephants before being floated to the sawmills. Other monsoon

[*Popper*

A Burmese Teak Forest

The elephant is *ounging*, i.e. rolling logs. With its tusks it can move six tons of teak.

products include millets in the drier areas; coconut palms, sugar-cane, and tropical fruits on wet coastal plains; and spices for seasoning, vegetable oils for fats, and indigo for dyes. Wheat grows only as a winter crop in the cooler and drier parts.

In their manifold peoples, languages, and religions monsoon lands show the same tropical abundance as in their fauna and flora. Underfed, diseased, and ignorant, India's millions live barely above starvation-level. Moreover, while soil erosion diminishes food supplies, population figures continue to rise in an alarming way, for the death-rate, though still very high, is declining because of medical skill and better sanitation.

Entirely different languages are spoken, and to an Indian from the north the speech of one from the south is quite unintelligible. Most Indians live in villages or small country towns on an income of a few shillings a week. Their straw-thatched hovels are made of bamboo, sun-dried mud, or—incongruous sign of the white man's civilisation—of flattened petrol tins. Enticed by

226

[*Popper*

An Indian spinning Artificial Silk
India is now an important manufacturing country.

higher wages to be earned in factory and mine, many have flocked into cities and mining towns, for India and Pakistan are undergoing an Industrial Revolution, though not nearly so rapidly as Britain did some 150 years ago. Bombay manufactures cotton cloth and Calcutta has jute mills. India is a leading producer of manganese, much of which goes to the Tata steelworks of Jamshedpur, one of the largest in the world. In Burma, too, there are valuable minerals, particularly rubies and oil.

During their long rule in India the British did much good work. Schools, hospitals, railways, factories, and dams and canals for irrigation were built. A great deal, however, remains to be done to raise the appallingly low standard of living of the illiterate ryot.

Monsoon North Australia.—Across the Indian Ocean from the crowded lands of South-east Asia lie the empty spaces of Northern Australia. Many Asians would welcome the oppor-

227

tunity to people this great tropical void, in spite of the fact that much of it is infertile and would attract none but the poorest settlers.

Australians, however, do not relish the prospects of a rapidly increasing Asian population overrunning their continent and competing with them in farm, factory, and mine. Coloured peoples are therefore allowed to enter the country only if they can pass a dictation test in any language chosen by the immigration officials. This method automatically excludes the unwanted immigrant. At the same time, it both safeguards the Australian from any possible accusation of racial prejudice and saves the Asian from "loss of face"—a matter of no little weight with peoples like the Chinese and Japanese. Europeans, however, are spared this ordeal, for Australia needs millions of white settlers to develop to the full her varied resources. Although the continent is twenty-five times as big as the British Isles, its population is little greater than that of London.

The emptiness of monsoon North Australia is emphasised by the small size of Darwin, the most important town. This port and route centre is no bigger than a little English market town. Along the north-western coast houses are sometimes chained down to resist the furious *willy-willies*, gales which destroy piers and jetties.

Abyssinia.—Mountainous Abyssinia, like South-east Asia and North Australia, in summer receives monsoon winds from the Indian Ocean (page 102). The Blue Nile, swollen by heavy relief rains, pours down in flood into the Anglo-Egyptian Sudan to join the White Nile. The flooded Nile then flows northwards across the desert, transforming it into the long, narrow, fertile, and densely populated oasis of Egypt. Whoever, therefore, controls Abyssinia controls the life-line of Egypt.

On their mountain-slopes nomadic Abyssinian herdsmen rear cattle, sheep, and sure-footed goats. Abyssinia was probably the home of the coffee tree, which is named after the province of Kaffa. The people live in circular stone huts, thatched with grass. Roads are rocky and rare, and goods are carried over mountain tracks by pack-animals. Nevertheless, civilisation has

[*Fox*

Summer Floods in Abyssinia

These Abyssinians are crossing a flooded river. The floods are caused by heavy monsoon rains in summer.

penetrated into these mountain fastnesses via the railway that links Addis Ababa, the capital, with the Gulf of Aden.

EXERCISES

1. Both India and North Australia have a monsoon climate, yet the former is overcrowded while the latter is practically empty. Account for the great difference in density of population between these two regions.

2. Explain the meaning behind each of the following statements about India:

(*a*) "No place in India seems to be less than three hundred miles from any other place; the longer journeys have to be measured in thousands."

(*b*) "Everything connected with India is on a scale which dwarfs the geographical proportions of Europe and makes them look almost ridiculous."

(*c*) "Civil prisoners in India always put on weight."

(*d*) "To kill a cow is one of the worst of sins."

(*e*) "The villagers use dirty water for drinking and cooking and regard the diseases which result as part of their fate."

3. In what ways did India benefit from British rule?

CHAPTER XXIV

The Hot Belt—2. Tropical Maritime Margins

WORLD POSITION AND CLIMATE

HEMMED in between eastern warm temperate and equatorial regions, tropical maritime lands lie along the eastern margins of continents from approximately 5° to 25° in each hemisphere (Fig. 73).

As the word "maritime" in their name implies, these regions show in their climate the influence of the sea. "Summer" and "winter" here have little meaning. The very small annual range of temperature (Fig. 75) is due partly to the persistently high sun-angle at noon, but especially to the ever-present control which the ocean has over the land. This maritime influence operates by means of on-shore trade winds, which over most of these regions blow at all seasons.

By their long passage over the oceans and by their gain in heat as they blow equatorwards the trades are enabled to absorb more and more water-vapour. On reaching land they are therefore both warm and wet. Over the sea and over plains trade winds do not readily part with their moisture, for condensation is checked by a continuous rise in temperature as they advance towards the Doldrums. High moun-

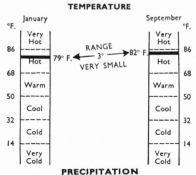

tation : GEORGETOWN (British Guiana)

TEMPERATURE

RANGE VERY SMALL 79° F. ← 3° → 82° F.

PRECIPITATION

Annual Amount : Very heavy (87·4″)
Seasonal Distribution : All seasons

Nov. to April 49%   51% May to Oct.

(Winter and Summer cannot be distinguished)

On-shore trades at all seasons

Fig. 75.

tains, however, fringe most tropical eastern margins and often face the ocean in steep scarps. Against these barriers the moist on-shore trades rise, are cooled, and yield heavy relief rains at all seasons (Fig. 75). The rains are especially heavy in "summer", when the winds are drawn ashore even more strongly than usual by the low pressure over the heated interiors of the continents.

No tropical maritime climate is experienced in dry North-east Africa. Here, with Asia instead of an ocean to the east, the trades have little opportunity to absorb water-vapour.

NATURAL VEGETATION AND ANIMALS

PERSISTENT heat and heavy, all-seasonal rains lead to the growth of evergreen rain forests of hardwood trees. These tropical jungles cannot be distinguished from equatorial forests, a description of which appears in a later chapter.

Monkeys, snakes, and birds of brightly coloured plumage abound. In its body the monkey is ideally designed for life in trees. Its limbs, unlike those of animals that remain on the ground, end in paws with four separate digits facing an opposable thumb, so that it can grasp branches and pick up food. Ball and socket fittings whereby its limbs rotate freely in all directions enable the monkey to swing from branches, an activity also aided, in some monkeys, by a prehensile, or grasping, tail.

MAN IN TROPICAL MARITIME MARGINS

WHERE he has cleared the forests man grows sugar-cane, tobacco, rice, pineapples, bananas, cacao, and coffee.

Caribbean Banana and Sugar-cane Plantations.—The West Indies and the Caribbean coastlands of South and Central America are famed for their banana and sugar-cane plantations.

Bananas are picked while green and are stacked in specially equipped ships, to ripen during the voyage to overseas markets. The tall sugar-cane is cut down by Negroes. Long ago their ancestors were brought to the New World from Africa, to toil as slaves for white plantation-owners, for in the damp, oppressive heat of this climate the white man himself finds it difficult to undertake strenuous manual work. The cane is hauled by

[*Tate and Lyle*

Cutting Sugar-cane on a Plantation in the West Indies
This picture gives a good idea of the height of sugar-cane.

bullock cart or by light railway to the factory, where its juice is converted into syrup from which sugar crystallises.

For centuries the prosperity of the West Indies has depended largely upon the export of sugar, but low selling prices have led to widespread poverty. In recent years British taxpayers have spent millions of pounds in raising the low standard of living here and in other British colonies. Moreover, it is proposed to relieve the pressure of population in these overcrowded islands by an overspill into British Honduras and British Guiana, both of which, if roads were built and forests cleared, could absorb thousands of immigrants.

[*Popper*

Picking Coffee Cherries on a Brazilian Fazenda
The pickers use little ladders to reach the topmost coffee cherries or berries.

Brazilian Coffee Fazendas.—Out of every ten cups of coffee made in the world, Brazil provides about seven. The coffee tree is planted in rich, red, volcanic soil on the Brazilian Highlands around São Paulo, the world's greatest coffee market. When young, it is shaded from strong sunlight by taller plants like the banana. In the heat of tropical lowlands fruits ripen at any time of the year, but in highland *fazendas,* as Brazilians call their farms, the cooler weather causes all the coffee berries to ripen simultaneously instead of at irregular intervals. In this way a considerable saving of time and labour in gathering the harvest is possible.

Along the north-east "shoulder" of Brazil Negroes grow sugar-cane and cacao, but production is seriously hampered by the old-fashioned methods of farming practised by these descendants of slaves.

Mahogany "Scouts" in British Honduras.—Man has raided the great stores of hardwoods in some tropical maritime forests. British Honduras, for instance, has for long been noted for its mahogany trade. Its hot-house atmosphere encourages plants to grow in a confusing maze of hundreds of intermingled species, and to find any particular type is somewhat like looking for a needle in a haystack. A "scout" therefore climbs a tall tree, and from this crow's nest maps the exact whereabouts of mahogany trees. If big enough, these are felled, sawn into logs, and hauled on tractor-drawn trucks or on light railways to the rivers, down which they float to the ships that carry them overseas.

From the sapodilla tree of these same forests oozes chicle, from which Americans make their chewing-gum.

Tropical Agriculture and the White Australia Policy.—Pine-apples, bananas, cotton, and especially sugar-cane are grown along the coastal plain of North-eastern Australia. Here, in Queensland, a great effort has been made to help to solve Britain's food shortage by planting maize as pig-food.

Australia's firm "No" to Asian immigrants means that Queensland's white sugar-planters must face serious competition from other tropical lands where cheap coloured labour is available. Home-grown sugar can be sold profitably in Australia only if a high tax is put upon imported supplies, whose selling price is thus forced up until it no longer undercuts that of the Australian crop. To maintain the policy of keeping coloured peoples out of her homeland, the Australian housewife therefore pays more for her sugar and for all home-produced manufactured foods in which sugar is an ingredient.

Mining.—The white man is just as eager to tap mineral wealth in these lands of heat and rain as he is where cold or drought prevails. Everywhere save in Australia, however, he employs coloured labourers to serve his purpose.

From British and Dutch Guiana bauxite is exported to Eastern Canada and the United States, where cheaply-produced electricity smelts it into aluminium. Trinidad's celebrated pitch lake yields asphalt for British roads, while Cuba's iron ore feeds blast-furnaces along America's industrialised north-east coast.

[*Shell*

A Survey Party in a Tropical Rain Forest

Preparations are here being made to map a forested area of Venezuela in a search for oil. What evidence is shown of the problems to be faced in the development of tropical forest lands?

Venezuela ranks third among the world's oil-producers. Queensland's Mount Morgan, a "mountain of copper with a cap of gold", was for long Australia's richest mine, but in his eagerness to exploit its wealth man has levelled it practically to the ground.

EXERCISES

1. The effect of the ocean upon climate in both tropical maritime and temperate maritime regions is great. On an outline map indicate by different shadings the world position of each of these climatic regions. Compare and contrast them in respect of (*a*) the means by which the ocean's influence is brought to bear upon them, (*b*) their climatic characteristics, and (*c*) their natural vegetation.

2. Explain why Negro slaves were shipped from Africa to the tropical plantations of the New World. How did the slave trade assist the growth of Liverpool and Bristol? What problems remain in various parts of the New World as results of Negro slavery in former times?

3. "History shows that, sooner or later, rich and sparsely populated territories must be defended." Arrange a debate on the advantages, disadvantages, and (as implied in the quotation) dangers of the White Policy of "Australia for the Australians".

CHAPTER XXV

The Hot Belt—3. Hot Deserts

WORLD POSITION AND CLIMATE

LYING roughly opposite to the tropical maritime eastern margins of continents are the hot deserts of their western margins (Figs. 73 and 77A). No greater contrasts could be imagined than those which exist between these opposite sides.

	Tropical Maritime	Hot Deserts
Prevailing winds	Moist, on-shore trades	Dry, off-shore trades.
Oceanic influence	Marked	Negligible (except along coasts).
Rainfall	Heavy, all seasons	Very light.
Natural vegetation	Thick forest	Sparse, thorny plants.

On the western sides of the southern continents are the Atacama and Peruvian Desert of South America, the Kalahari Desert of South Africa, and the desert zone which reaches from the "dead heart" of Australia to the Indian Ocean.

In the northern hemisphere a large arid area covers the South-western States of America and continues into the north-west of Mexico. It includes Death Valley and the Mohave, Colorado, Sonora, Gila, and Painted Deserts. In North Africa, however, the Sahara Desert is not confined to western margins. It extends across the whole continent from the Atlantic Ocean to the Red Sea. Then, under other names, it continues via Arabia to Iraq and finally merges into the temperate deserts of Central Asia. The Sahara proper is nearly as big as the United States, or thirty times larger than the British Isles.

Temperature.—In summer the noonday sun burns "all in a hot and copper sky", and drives up the thermometer to break world

records. Day-time temperatures of well over 100° F. in
the shade are of common
occurrence, while Tripoli, in
North Africa, holds the
record maximum temperature
of 136° F.

Describing the desert of
South Palestine, A. W. King-
lake writes: "The heat grew
fierce. . . . Hour by hour I
advanced, and still there was
the same, and the same, and
the same—the same circle of
flaming sky—the same circle

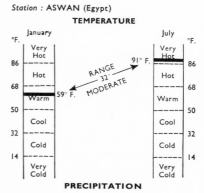

Station : ASWAN (Egypt)

TEMPERATURE

PRECIPITATION

Annual Amount : Very light (0″)
Seasonal Distribution : Drought at all seasons
Dry, overland and off-shore trades

Fig. 76.

of sand still glaring with light and fire. Over all the heaven
above, over all the earth beneath, there was no visible power
that could balk the fierce will of the Sun. . . . From pole to
pole, and from the East to the West, he brandished his fiery
sceptre as though he had usurped all Heaven and Earth."

Winter days, however, are pleasantly warm. Winter can
generally be distinguished from summer far more easily here
than in any other climate of the hot belt, the annual range
being moderate or even great (Fig. 76).

After sunset the store of heat hoarded up by the earth during
daylight hours is quickly radiated away into a clear, star-
bespangled sky. Nights are consequently cool and, in winter,
frosty. In fact, the daily range often outstrips the annual range
of temperature.

Precipitation.—The annual rainfall is very light. The trade winds,
here off-shore, have already parted with their moisture during
their long overland journey. In fact, where their path is barred by
high mountains like the Andes they do not even reach the deserts.
Occasionally summer heat draws ashore sea breezes. Such on-
shore winds should bring relief rains. They have, however, been
chilled by their passage over the cold ocean currents that wash
these desert shores. Once ashore, their temperature rises as they

are warmed by the heated land, and under such circumstances the condensation of water-vapour is impossible (page 106).

Such precipitation as does occur comes either as heavy dew during the chilly nights or as very rare but severe convectional thunderstorms. These heavy rains convert sand into quicksand and *wadis*, or dried-up watercourses, into torrents. Strange as it may seem, travellers have been drowned in the Sahara Desert.

Natural Vegetation and Animals

In hot deserts natural vegetation is adapted to resist drought far more than anywhere else. Some deserts form bald patches on the earth's surface. Others are thinly covered with thorny plants like the cactus.

If watered, desert soil is very fruitful, for there is little rain to dissolve and wash away its fertile salts. Seeds lie dormant for years, awaiting a life-giving shower. After rain the desert may "blossom as the rose", but its carpet of gay flowers quickly withers under a hot sun. The seeds are shed, and the cycle of seeding, growth, and death is completed all too soon.

Wherever underground water approaches or reaches the surface oases dot the parched landscape with green "islands". An oasis is not necessarily a mere pond bordered by vegetation; the larger ones contain towns set amid cultivated fields and groves of date palms.

Desert animals, like plants, are adapted to their arid environment. Some insects never drink, while other creatures absorb dew through the skin. Many animals are coloured to match their background and so, by camouflaging themselves, avoid their enemies. Some keep cool in caves or in burrows. Where caves are scarce an animal that hunts by day may take turns with a nocturnal prowler in sharing the same lair. Some have long hind legs, like the jerboa, or desert rat, which in speed can rival a race-horse.

The camel has been well named "the ship of the desert". Its spreading feet support it over yielding sands, while its eyes and nostrils are protected from wind-blown sand. Fatty food is stored in its hump and water in its stomach. For three days at a stretch it can go without water, while thorny plants cannot

[C. P. Mountford

An Australian Aborigine returns from a Hunt

This aborigine is carrying a kangaroo which he has killed with his ten-foot hunting spear. He is followed by his dingo. Normally the dingo is a wild dog, but those kept by the aborigines have been tamed. Because it kills sheep the dingo is shot by "doggers", who earn a living from the bounty paid by the Australian Government for each scalp.

gash its hard mouth. It supplies man with milk, meat, and hair for tents and clothing.

To carry stores to mining camps and sheep farms, the camel was introduced as a pack-animal into the dry parts of America and Australia. In the Sahara camel caravans toil from water-hole to water-hole and make trade possible between widely separated areas. In this way desert salt is exchanged for the hides, cotton, and grain of savanna lands, while dates are carried from oases to the nearest railhead.

MAN IN HOT DESERTS

JUDGED by their way of life, desert-dwellers fall into four groups: Primitive hunters and collectors; Nomadic herdsmen; Agriculturists, or crop-growers; and Miners.

Primitive Hunters and Collectors.—For their food and their trifling needs of clothing and shelter primitive hunters rely upon whatever animals they can catch and whatever plants they can find.

The Bushman of the Kalahari Desert speaks in guttural clicks and cannot count beyond two or three. A skilful tracker, he kills his prey with a wooden club or by a bow and poisoned arrow. He is an expert at finding underground water, which he sucks up through a reed and stores in empty ostrich shells.

Like the Bushman, the Australian Blackfellow, or aborigine, is an excellent tracker. If he were not he would starve. So fine is his bushcraft that occasionally the police employ him to hunt down criminals. To him nothing is unpalatable—snakes, lizards, kangaroos, frogs, seeds, roots, wild honey, and cooked grubs are all eaten with gusto. Formerly some aborigines were cannibals.

Both the Bushman and the Blackfellow are gradually dying out as the white man's civilisation and its diseases invade their homelands. Illnesses from which a European usually recovers often prove fatal to primitive peoples who are unaccustomed to them.

Nomadic Herdsmen.—Nomadic pastoralists, or herdsmen, like the Arabian Bedouin and the Saharan Tuareg roam the deserts in search of pastures for their sheep, goats, horses, and camels.

Like many inhabitants of wide open spaces, the sunburnt nomad is tall and lean, with a hawk-like nose and high cheekbones. Normally he is more healthy than the town-dweller in the oasis, where overcrowding, impure water supplies, and insanitary houses lead to the outbreak of disease.

His white clothing reflects the sun's rays and so keeps him cool. Over a cotton robe is flung a cloak of camel hair, while a cloth head-dress, bound with a circle of cord, protects the head from heat by day and from cold by night.

These nomads live in tents of black goat's-hair cloth, which are low-pitched to withstand violent sandstorms. Whereas we pull aside our curtains to welcome warm sunlight, the Arab seeks shade and so always lowers the flap on the sunny side of his tent, although the remaining walls are raised to admit fresh air. Desert pastoralists, like nomadic steppe herdsmen, own

Nomads in the Arabian Desert   [*Esso*

These Bedouin nomads are watering their sheep, goats, and camels at a well originally drilled by an American oil company. From the trough a pipe-line leads to a diesel pump at the well. The low-pitched tents withstand sandstorms.

few personal possessions. These are easily portable, consisting of tents, rugs, weapons, and household pots and pans. The nomad's wealth lies in his flocks and herds.

The main meal of dates, bread, and coffee or milk is eaten after sunset before a fire of brushwood or camel dung. Rice with meat provides an occasional luxury. Arabian hospitality is proverbial and the stranger is always sure of a welcome, for the Arab knows want and poverty too well to refuse to share his meal with others.

His constant struggle to wrest a living from the desert makes the nomad warlike. In times of scarcity nomads formerly plundered well-stocked camel caravans or agricultural oases, but the French Government has brought law and order to the Sahara, and breaches of the peace are severely punished. Of all desert-dwellers the Tuareg "People of the Veil" are the most warlike. The Tuareg wears a black tunic over cotton trousers and carries a dagger, a lance, and a sword strangely like that of the Crusaders of old. His head, all but his eyes and the tip of his nose, is swarthed in a black veil. Probably originating as a protection against the dreaded sandstorm, this veil is worn religiously until death and the Tuareg permits no one to see his face unveiled.

The Sahara has not always been as big as it now is. Many an African city that flourished during the Roman Empire is now buried beneath desert sands. The nomad has helped to turn fertile land into wastes by grubbing up soil-binding trees for fuel, so enabling sun and wind to play havoc with the loosened soil. His greedy goats are equally to blame, for they rip grass up by its roots and feed on saplings; they have been deservedly nicknamed "black locusts". Nowadays laws strictly forbid the nomad to tear up trees and to pasture his animals where sand-binding grasses are holding the desert in check.

In hot deserts, as in other climatic regions, civilisation is making its mark. In North Africa nomads are encouraged to improve their breeds of sheep and camels. In some parts annual markets or fairs are held, just as we have county agricultural shows. The motor lorry and the railway may in time make camel caravans as out-of-date as the pack-horse became when the Industrial Revolution brought canals and railways to Britain. Some large Saharan oases are already linked by rail to Mediterranean ports, and the French Government has planned a railway to cross the Sahara to Timbuktu, a market-town where desert merges into savanna lands.

Agriculturists in Oases.—Two problems face those who, in oases, live the settled life of agriculturists, as the tillers of the soil are called. The first is to get water, either from rivers or from underground sources. Secondly, this precious water must be led to crops with as little wastage as possible.

Irrigation from Rivers.—In some deserts Nature solves the problem of water shortage by providing rivers for man's use. In Iraq and Egypt water is lifted from river and irrigation canal by the *shaduf*, a lever with a weight at one end and a bucket at the other, or is scooped up by the *sakia*, a primitive water-wheel worked by a donkey, buffalo, or camel. Nowadays these and other primitive methods of raising water are being replaced by mechanical pumps. These are owned by companies, and the peasants pay for the water so obtained.

Basin and Perennial Irrigation.—When the Nile is in flood a method of watering crops known as *basin*, or *flood*, irrigation is practised in Egypt. The land is divided into small earthen-walled fields called basins, which stretch outwards from the river to the limits to which flood-waters can be led. By September summer floods from monsoon Abyssinia reach Egypt, and the flood-waters pour into the basins through breaches in their earthen ramparts. Eventually they ebb from the fields, leaving in their wake a fertile covering of silt. Immediately the *fellahin* (Arabic=tillers of the soil) plough the moist soil and plant wheat, clover, onions, and beans. Ripening beneath a warm winter sun, these crops by February await harvest. As Shakespeare says:

> *The higher Nilus swells*
> *The more it promises: as it ebbs, the seedsman*
> *Upon the slime and ooze scatters his grain,*
> *And shortly comes to harvest.*

Cotton and sugar-cane demand a longer growing-season and more heat than do these winter crops. They must be planted in spring, well before flood-time. Water for their growth is therefore stored from the floods of the previous autumn. Ponded back in reservoirs behind dams which were built by British

The Nile is the Life-line of Egypt

In this aerial view of the strip of irrigated land that borders the Nile note the sudden change from fertile fields to desert.

engineers, the water is drawn off as required. By this method of *perennial* irrigation thirsty crops can be moistened at any season. The word "perennial" means "throughout the year". Every fortnight during the summer the irrigation channels between the rows of cotton bushes must be refilled. The supply of water to each grower is strictly rationed. Sluices allow surplus flood-waters to pass through the dams so that farmers much farther downstream will not go short. Perennial is much more widely practised than basin irrigation, which is comparatively unimportant.

Most Egyptian farms are little bigger than a football field. The hard-working fellahin, toiling with their hands on tiny fields, resemble Asian peasants rather than the farmers of Canada, America, and Russia, with their large-scale, mechanised agriculture. Their villages are strung along the edge of the long strip of irrigated land that fringes the Nile, or on islands of slightly higher ground above the reach of flood-waters. Consequently the whole of the fruitful zone is devoted to food production.

The change from the emerald-green carpet of crops to the bare yellow floor of the desert is both sudden and startling.

Describing the view from the top of the Pyramids, Mark Twain writes: "On the one hand, a mighty sea of yellow sand stretched away towards the ends of the earth, solemn, silent, shorn of vegetation, its solitude uncheered by any form of creature life; on the other, the Eden of Egypt was spread below us—a broad green floor, cloven by the sinuous river, dotted with villages, its vast distances measured and marked by the diminishing stature of receding clusters of palms."

In Iraq high tides in the Persian Gulf are made to serve man. The tidal current moving upstream dams back river-water until it spills over into the honeycomb of irrigation channels covering the Tigris-Euphrates delta. This region produces 80 per cent of the world's dates. Here, according to local legends, was the site of the Garden of Eden—certainly no greater contrast can be imagined than that between the cool paradise of the shady date-groves and the burning heat of the desert beyond.

In the "little Egypts" of Peru cotton and sugar-cane are watered by some sixty streams fed by melting Andean snows. Irrigation is also very important in the dry South-western States of America, where the waters of the Colorado River are held back by the famous Boulder Dam.

Irrigation from Underground Water.—In some deserts like Arabia a quarter of a century may pass without a shower, while the total rainfall during a "good" year may last for one or two hours. No permanent rivers flow through such wastes and man must therefore delve underground to find water.

Sand-pit Oases of the Erg.—In the sandy parts of the Sahara, known as the *erg*, man digs hollows deep enough to bring subterranean water supplies within reach of the moisture-seeking roots of his date palms. The date palm truly grows with "its roots in water and its head in fire". The ground in which it is grown is not owned as private property, but the tree itself is. Date palms are bought and sold, and in any one garden the trees may belong to several owners. Those around the edge of a hollow command higher prices than those in its centre, for their buyers can widen the sand-pit and increase their wealth by planting more trees.

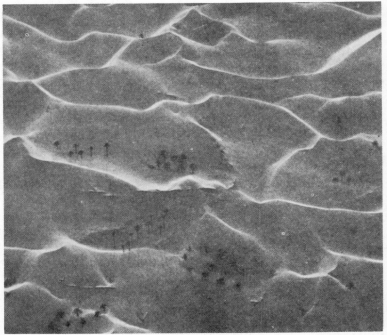

[*Emil Brunner*

A Saharan Oasis overwhelmed by Sand

In this oasis only the crowns of date palms and the tops of walls show above the wind-blown sand.

Wherever a hollow, either man-made or natural, actually pierces water-bearing rocks, springs bubble out and create small lakes. Ditches guide water from these pools to the fields, and sluices measure out to each farmer his correct ration of precious water.

The inhabitants of these sand-pit oases, which are about thirty feet deep, must continually resist the invasion of their sunken gardens by wind-blown sand. Ceaselessly they toil at the endless task of carrying up the slopes of the oasis sand-filled baskets perched upon their heads or on the backs of donkeys.

Wells in the Hammada.—In the *hammada*, or rocky areas of the Sahara, deep wells are sunk to tap underground water. Above the mouth of the well a pulley carries a rope fastened to a huge

246

[*Emil Brunner*

An Oasis Town in the Sahara Desert

Streets are narrow, thus providing shade and leaving room for each building to enclose a central courtyard. Domed roofs reduce heat indoors. On the outskirts are "sand-pit" date palmeries.

leather bucket. A long slope leads up to the well, and down this inclined plane trots a donkey, camel, or a man, hauling at the rope and thus raising water with the minimum of effort. The deeper the well, the longer is the slope. Day and night this work continues without ceasing, for the owners of those valuable wells that never run dry sell to others, at so much an hour, the right to draw water.

Occasionally wadis are barricaded to hold up floods after convectional storms.

Watering the thirsty land from wells is a never-ending task, and the water must be used wisely. To prevent seepage into the ground irrigation ditches are lined with earthenware tiles or concrete, and to reduce evaporation water is sometimes led along underground channels.

In some oases the richer people pass the summer in their country houses. These are built in the cool shade of irrigated gardens, planted well outside of the towns. Here a miraculous transformation scene is brought about by the power of water, for these gardens, although surrounded by desert wastes, are like man-made copies of the luxuriant vegetation of an equatorial forest. Figs, apricots, vines, and pomegranates abound, while wheat, barley, and beans flourish between rows of date palms. The owners return to spend winter in the towns, which in many cases crown hill-tops and so in former times were the more easily defended against marauding nomads.

Desert Towns.—Building material in the desert differs from place to place. In Egypt, where stone is scarce, houses are built of dried mud or clay, with a flat straw and mud roof supported on two beams. Sometimes they are entirely roofless. In other desert regions stone houses appear, their roofs being either flat or else domed like bee-hives. Furnishings are generally meagre, partly on account of the poverty of the householders and partly because so much time is spent out of doors that a richly equipped interior would be a waste of money.

Sometimes the family seeks relief from summer heat in cool underground rooms. In Baghdad, for instance, the *serdab*, a cellar six feet below street-level, sends up ventilation shafts whose cowls catch the *shamal*, a local wind, and pass it down as a cooling draught. The streets of towns in hot deserts, and, indeed, throughout the hot belt, are generally very narrow. They certainly provide welcome shade but are unfit for heavy modern traffic.

Miners in Hot Deserts.—Not even the heat and drought of the desert can daunt the white man in his quest for useful or precious minerals.

Deserts in Western Australia and the south-west of the United States yield gold; Algerian iron ore feeds British blast-furnaces; and the Atacama Desert of North Chile ranks second only to America in copper mining. Preserved by the dry atmosphere, fertile salts abound, e.g. phosphates in the Sahara, nitrates in North Chile, and borax in America's Death Valley. Coal is now mined in the Sahara, while Britain's political interest in the Near East is partly to be explained by the rich oil-wells of Iraq and Iran, sunk by British engineers and paid for by British money.

The days of the lone prospector, seeking a fortune in out-of-the-way places, are practically over, and many a derelict "ghost" town bears silent witness to former prosperity. Nowadays minerals are produced by big companies which can afford expensive mining machinery. Millions of pounds are sometimes spent in constructing roads, railways, or pipe-lines from oil-well or mine to port, and in pumping water to desert mining towns from areas blessed with a heavier rainfall.

Recently the white man has found new uses for unpopulated deserts as ideal testing-grounds for the rockets, supersonic aeroplanes, and atomic bombs with which, if he so wills, he can shatter his own hard-won civilisation.

A peep into the future might even show us factories in deserts like the Sahara, worked by solar engines trapping power from the intensely bright rays of the sun.

EXERCISES

1. On an outline map of the world shade and name the hot deserts. Why is the Sahara the only one to reach the eastern margins of a continent? Describe and account for the climate and natural vegetation of hot deserts.

2. Compare and contrast the mode of life of the Bedouin of the desert with that of the Khirgiz of the steppes.

3. The following figures refer to Egypt:

Total area	.	.	.	.	393,000 square miles
Cultivated area .	.	.	.	13,100 ,, ,,
Density of population in settled area	.	1,300 per square mile.

(*a*) What percentage of the total area is cultivated? Why is this percentage so low? (*b*) Why is the density of population so great in the settled area?

CHAPTER XXVI

The Hot Belt—4. Tropical Continental Lands

WORLD POSITION AND CLIMATE

IT should be obvious that the almost rainless deserts of the western margins of continents in trade wind latitudes cannot suddenly be transformed eastwards into the rainy tropical maritime eastern margins and equatorwards into the equally rainy equatorial regions (Fig. 77A). There is bound to be an intermediate transitional belt of moderate rainfall, shown in Fig. 77B, which receives more rain than the deserts on the one hand, but less than tropical maritime and equatorial lands on the other.

Tropical continental regions (Figs. 73 and 77A) provide this

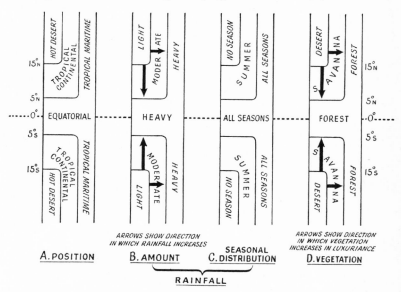

Fig. 77.—Position, Rainfall, and Vegetation of the Regions of the Hot Belt

half-way stage. As their name suggests, they do not come under oceanic influences as tropical maritime regions do. This kind of climate is sometimes called Sudanese, for it is typical of the Sudan in North Africa.

The orderly patterns shown in Fig. 77, however, are to be seen clearly only in the southern hemisphere, and even here there are flaws. North of the Equator the irregular distribution of continents and oceans drastically upsets this ideal plan.

Temperature.—A very hot season is followed by a somewhat cooler season (Fig. 78). Here, as in much of the hot belt, it is really better to distinguish seasons as *rainy* versus *dry*, rather than as *summer* versus *winter*. The annual range of temperature is certainly more noticeable than in tropical maritime and equatorial lands, but it is not so great as in hot deserts. Often the daily range between the burning heat of the day and the coolness of the night is far greater than the mean annual range.

Precipitation.—In passing through these regions, the more one retreats from hot deserts and advances towards tropical maritime and equatorial lands, the heavier becomes the rainfall, as shown by the directions of the arrows in Fig. 77B. The rain in tropical continental lands, as in their temperate continental namesakes, is convectional and comes chiefly in "summer" if, for convenience, we continue to call the hotter season by this name (Figs. 77C and 78).

Those parts immediately bordering the equatorial belt receive their convectional

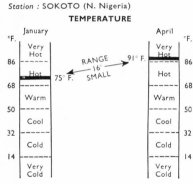

Station : SOKOTO (N. Nigeria)

TEMPERATURE

Note that the hottest month is not July, but April, i.e. before skies become covered by rain-clouds

PRECIPITATION

Annual Amount : Moderate (25·4″)
Seasonal Distribution : "Summer" rain,
"Winter" drought

WINTER 1%   99% SUMMER
(*Dry* and *rainy* are better names for the seasons)

Dry, overland and   Convectional
off-shore trades   thunderstorms

Fig. 78.

251

thunderstorms from the low-pressure calms of the Doldrums when, lagging slowly in the wake of the migrating overhead sun, they visit these areas in "summer". Those parts that lie sandwiched between hot deserts and tropical maritime regions are likewise drenched by "summer" convectional storms, but not in this case from the Doldrums, which do not stray thus far from their equatorial haunts. A British District Officer has described the onset of the rainy season thus:

"Lightning flickering below the horizon and the daily gathering of cloud promise rain before the tour is over. It also gives us all an opportunity to discuss the weather—a pastime which I, as an exiled Englishman, sadly miss. . . . The storm came suddenly. When I closed my eyes at one o'clock the only occupants of the sky were several hundred flat-bottomed little silver clouds. Ten minutes later the trees were rocking in the gale, and rain was driving down so hard that sheets of spray were beating up off the ground . . . The roof of the hut stood up to the onslaught for about two minutes. Then the water came pouring through" (K. Bradley).

This season of thunderstorms gives way to a season of drought, when all tropical continental lands come under the sway of trade winds. By the time they reach here these winds are dry, having yielded up their moisture as relief rains in crossing the tropical maritime highlands of eastern margins.

Over West Africa the north-east trade wind blows out from the Sahara and is known locally as the *harmattan*. It blows in the cooler season and occasionally reaches the Guinea Coast. So intensely dry is it that one's skin becomes parched and finger-nails crack. Tree-trunks split, while a dismal pall of choking grey dust blots out the sun and mantles the whole countryside. Although it is really a warm wind, its dryness so aids the evaporation of perspiration that one feels refreshingly cool. Along the Guinea Coast this welcome relief from the normal oppressive heat induces the Negro to call the harmattan "the doctor". Farther north, however, its extreme dryness causes an uncomfortably great daily range of temperature that partly accounts for the high death-rate from pneumonia among the lightly clad Hausa peoples in their overcrowded cities.

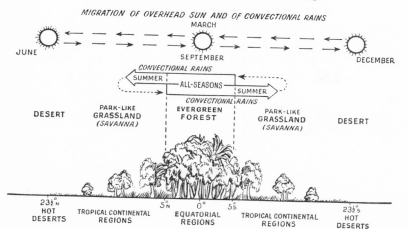

Fig. 79.—Forest, Grassland, and Desert in the Hot Belt

Note how the belt of convectional rain lags behind the migrating overhead sun. This transition from equatorial forest to hot desert through savanna lands occurs along the western margins and in the interior of continents. Along eastern margins (except in N.E. Africa) forests are unbroken.

NATURAL VEGETATION AND ANIMALS

THE hot rainy season encourages the rapid growth of very tall grasses, e.g. elephant grass, which sometimes towers to a height of ten feet or more. Often this grassy carpet does not completely cover the ground, but shows holes where bare earth separates isolated clumps of grass. Scattered umbrella-shaped trees, well adapted to survive the dry season, give these *savannas* a park-like appearance. Fringing the rivers are continuous strips of gallery forest. Characteristic among the trees of savanna park-lands is the baobab, whose swollen trunk is a natural water-butt. In the manifold uses to which it is put the baobab rivals the bamboo of monsoon Asia.

In vegetation, as in climate, savanna lands form a half-way, transitional stage between hot deserts and equatorial forests. Typical savanna scenery is best to be seen in the middle of tropical continental lands. Towards the deserts savannas thin out; the grass becomes stunted and wiry, and the trees dwindle into thorny bushes. Conversely, towards rainy tropical maritime and equatorial regions trees increase in height, variety, and numbers, crowding out the grasses until savanna merges into forest (Figs. 77D and 79).

In the dry season the withered grass dies, and in some areas it is burned by local tribes to make way for cultivation during the following rainy season. The trees shed their leaves to check loss of water by transpiration. The vivid contrast between the bare dry-season rags of tropical savannas and the rich green mantle in which they are clothed, almost overnight, when the rains come is well shown in the following passage:

"After four months of drought Nature seems altogether to have abandoned the struggle to keep up her good looks. There is no colour anywhere. We walked through country which had been blackened with recent fires. . . . The trees were either bare or hung about with patches of dead leaves, wrinkled and dry and as black as the ground beneath . . . But, oh, the smell of Africa after the first rain! Nowhere in the world is the scent of wet earth so pungent and so sweet. And then within three days the face of all the countryside has changed. Blackened *dambos*, brown hill-sides, yellow lawns—all are as green as English fields" (K. Bradley).

In Africa the east-west "corridors" of savannas on either side of the equatorial forest are connected by a broad savanna "passage" over the East African Plateau of Kenya, Uganda, and Tanganyika (Fig. 61). Being equatorial in latitude, these highlands should be forested. Their height, however, makes them cooler than the neighbouring lowlands and, since in equatorial latitudes heat leads to rain, the rainfall here is lighter than is usual and will not support dense forests.

Savannas provide an ideal home for big game, particularly in Africa. Long legs enable many animals to widen their horizon in these vast open spaces and to escape rapidly when danger threatens. Some, like the mottled giraffe and the striped zebra, are by Nature camouflaged to merge inconspicuously into the landscape. Towering nearly twenty feet high and with an eighteen-inch tongue, the giraffe can easily browse on the leaves of trees.

Herbivorous antelopes, zebras, giraffes, and gnus fall prey to carnivorous lions and leopards, with the hyena, jackal,

Tropical Savannas   [Fox

Against this Kenya background of tall grasses and scattered trees giraffes and other striped or spotted animals are difficult to detect.

and vulture following up as scavengers to clean up the carrion.

Ticks, termites, and tsetse flies unite to plague man. The cattle tick brings fever to his herds. The termite, commonly called the white ant, although it is not really an ant, gnaws away the woodwork of his dwellings and furniture. The tsetse fly spreads sleeping sickness amongst man himself and serious disease among his domestic animals. Where roads and railways are lacking in fly-infected areas, human porters replace pack-animals, which all too readily succumb to disease. Man has sometimes been forced to retreat before the invasions of these flies and to seek a home elsewhere. Another pest is the locust. Laying waste the countryside as if by fire, it consumes man's crops to the last stalk. In the swamps, ponds, and rivers live water snails that convey to man parasites which cause bilharzia. This enfeebling disease prevents millions of coloured people

[*Popper*

A Swarm of Locusts in Kenya

Locusts consume crops and pastures, and cause thousands of cattle to starve. In East Africa an anti-locust survey is maintained in the dry north, where many swarms originate.

throughout the hot belt from working to the best of their ability. In some parts the white man is stamping out the snails by dosing their moist haunts with copper sulphate. Another method of fighting the disease is to plant alongside the rivers certain trees whose leaves, when infused in the water, poison the snails. This practice is adopted along the canals of Egypt, where bilharzia is particularly common.

MAN IN TROPICAL CONTINENTAL REGIONS

THE inhabitants of savanna lands may be divided into: nomadic herdsmen; native agriculturists; white settlers; and miners.

Nomadic Herdsmen in African Savannas.—The numerous tribes of African herdsmen, although they differ in customs and

Cleaning Breeding-grounds of Tsetse Flies in Nigeria　[*C.O.I.*

Land clearance is one method of reducing sleeping sickness and nagana. Once the clearance has been made, maintenance work is undertaken by the community, each adult male giving four days' labour a year.

appearance, alike depend upon cattle and, to a less extent, upon goats. Wealth is judged by the size of the herd. Quantity comes before quality, and frequently far too many animals are reared upon the available pastures. Soil erosion follows this reckless overgrazing of grassland and huge areas, hopelessly ruined, have been abandoned.

In some parts the white man encourages the African to improve his breeds of cattle. From 15° N. to 15° S., over a vast area nearly forty times as big as the British Isles, the tsetse fly formerly held undisputed sway. When it bites cattle this pest passes into their blood the parasites that cause nagana, a disease fatal to all but the poorest breeds. Now, however, science has come to the rescue, for good-quality beasts can be inoculated against nagana with antrycide, a drug first discovered in British chemical laboratories. Unfortunately, this drug does not afford

[C.O.I.

A Test Herd of East African Cattle being injected with Antrycide
By such experiments the effectiveness of new drugs against disease can be determined. The treated cattle will be sent to an infected area.

a permanent safeguard and the problem of protecting cattle from the tsetse has by no means been fully solved.

To fight the fly itself powerful insecticides are used, while the shady brushwood in which it breeds, especially along gallery forests, is cleared and burned. Should these remedies prove successful, Africa may become one of the world's great beef-producers, although other diseases, e.g. rinderpest, must also be conquered before this hope can be fully realised. In Bechuana-land work has already been started on the establishment of ranches, together with a slaughter-house and freezing plant capable of handling nearly 100,000 cattle a year. Crops for cattle-fodder will be grown on 800 square miles of well-watered land.

That the prosperity of savanna nomads depends upon their cattle was well shown in 1870, when rinderpest almost exterminated the herds of the Masai people of East Africa and led to widespread starvation. The peaceful, humdrum existence of the agriculturist is scorned by the Masai nomad. His is the adventurous life of the wanderer, unhampered by cumbersome personal possessions. The bravest warriors used to protect the herds by hunting marauding lions, an adventurous pursuit now less necessary than formerly. They also stole cattle from other tribes, by whom they were greatly feared. Nowadays, however, British officials keep the peace, just as the French have subdued the equally ferocious Tuaregs of the Sahara.

Besides the milk and flesh of cattle and goats, the Masai occasionally consume blood. This luxury is drawn through a puncture pricked in the skin of the living cow. Although grain and vegetables are eaten by the women and elders, the young men disdain the fruits of agriculture.

At nightfall these nomads retire to their huts after seeing that their herds are penned within a thorn-hedge enclosure, safe from prowling lions. Made of grass and brushwood woven around a framework of branches, these flat-roofed and window-less huts are about six feet high and contain several rooms. The bedding consists of hide-covered mattresses of brushwood and grass. As in hot deserts, a brushwood fire gives comfort on chilly nights.

In West African savannas the nomadic Fulani roam with their cattle, sheep, goats, asses, horses, and camels, although a few have settled down to an agricultural way of life. Some are in a half-way stage, rearing animals but also growing crops. The men mind the herds while the women busy themselves with spinning and weaving and with household duties. Like the Masai, the Fulani at night safeguard their animals in a thorny stockade.

The Fulani provide a good example of the way in which, in various parts of the world, nomadic wanderers have conquered agricultural settlers. Early in the nineteenth century they sub-dued the Hausa cultivators of West Africa, and for a century they formed the ruling classes.

Their small circular huts with domed roofs are made of grass and brushwood but are not built to last long, for the Fulani move on when their animals have exhausted the local pastures.

Native Agriculturists in African Savannas.—Although methods of cultivation among primitive peoples are crude, agriculture generally calls for the exercise of more thought and foresight than does the rearing of animals. Because of these demands upon the brain, crop-growers are generally more advanced than animal-herders, although there are, of course, exceptions to this rule. Some African agriculturists are, indeed, highly civilised people, skilled in the manufacture of arms, leather goods, pottery, jewellery, and hand-woven cloth.

Mealies, as maize is called in parts of Africa, millet, tobacco, cotton, ground-nuts, and fruits like the banana are the leading crops in tropical continental regions. Agriculturists here either practise *subsistence farming* or grow *cash-crops*.

Subsistence Farming.—By this term is meant the cultivation of crops entirely for the grower's personal use and not for sale. The most primitive agriculturists burn down the grasses in the dry season and plant their seeds in good time to benefit from the heat and moisture of the rainy season. In West Africa, instead of each man tilling his own plot of ground, Negro *hoe-cultivators* co-operate in small groups, singing tribal songs as they scratch the soil with their short hoes. European agricultural experts are striving, not without success, to persuade the African to exchange his primitive hoe for the animal-drawn plough.

Within a few years the soil's fertility is exhausted, not only in feeding crops but also because of *leaching*, the process by which fertile salts near the surface are dissolved and carried well underground by heavy rains. Once the soil becomes barren, the people move on to continue the destruction of fertility elsewhere.

Africa cannot really afford to practise either this wasteful *shifting agriculture* of the crop-grower or the equally destructive overgrazing of the herdsman's cattle. With medical skill reducing the formerly high death-rate, the African population is doubling itself every thirty years, and food supplies should not be allowed to dwindle through man's misuse of Nature's bounty. In recent years, therefore, the white man has done much to make tribesmen realise the folly of these ruinous methods of farming.

Cash-crop Farming.—Cash-crop farmers, although they usually grow some of their own food, cultivate crops to sell either to neighbouring peoples or to customers overseas. With their profits they buy manufactured goods. Thus a natural "give and take" trade flows between temperate countries, which need tropical foodstuffs and raw materials, and tropical cash-crop growers, who want cloth, sewing-machines, and bicycles.

The Hausas of West African savannas are typical cash-crop farmers. Intelligent and skilled craftsmen, they weave cloth and work in glass, leather, and iron. In the rainy season they grow millet, maize, cotton, and ground-nuts. Some of the cotton is sold for export, while many a Sudanese Negro wears cloth made on Hausa hand-looms. The ground-nut plant buries its blossoms underground to produce nuts from which oil is crushed for the manufacture of margarine.

In some areas Hausas check the decline in soil fertility by practising a rotation of crops and by persuading Fulani herdsmen to keep cattle on land that needs manure. By such methods the land is kept in good heart, and shifting agriculture becomes unnecessary.

Hausas usually live near the rivers that supply them with water. In the dry season they tap underground water supplies, either by sinking wells or by digging below the dried-up beds of streams. Unlike most African peoples, they build towns surrounded by mud walls, within which flat-roofed houses are to be seen fringing narrow, shady streets. Long spouts, projecting well beyond the buildings, lead the overflow from convectional thunderstorms far away from the mud walls.

The Hausas have schools and some of them may even go to the new University College at Ibadan. Communications, however, are poor. Railways are almost non-existent and the earthen roads, parched and dusty in the dry season, are churned into quagmires by the rainy season's thunderstorms. It is no wonder that the European prefers to travel by aeroplane.

Formerly, frequent wars between cattlemen and cultivators turned the walled towns into forts. The British and French, however, brought peace to West Africa, as well as improved methods of farming. In Gambia British enterprise has cleared large tracts of bush so that grain may be grown to feed cattle, pigs, and poultry. The European adviser has also urged many villagers to give up their traditional custom of holding land as common property and to become private landowners.

White Settlers in Savanna Lands.—In Australian savannas the white man, like the African nomad, is a cattleman. Since Australian beef is exported, especially to Britain, great care is taken to rear cattle of good quality and to overcome disease.

The rancher lives in a wooden house with a wide, shady verandah and a roof of corrugated iron. Outside stands a tank for storing rain-water. The cattle roam freely over huge *stations*, or ranches, some of which are larger than Belgium. Over such vast distances a wireless transmitter, worked by a pedal, puts one in touch with one's neighbour or, in case of sickness, with the nearest doctor. Should a serious illness occur, a "flying doctor" takes the patient to hospital by aeroplane. The mail is also delivered by air, while children in the most remote districts are taught by free correspondence courses.

After the annual round-up the "fats", as cattle fit for slaughter are named, are driven by mounted stock-riders over wide grassy tracks called stock-routes, or "pads", to the nearest railhead. The slowly moving sea of cattle takes weeks or months to complete this long trek, which often amounts to hundreds of miles. Grazing as they go and watered from streams or from artesian wells, the cattle are kept in check by long stockwhips, for if they stampede they run off their meat. From the railhead they go by train to the ports where, after a fattening on maize, they are turned into chilled or frozen beef in freezing-factories. In North-west Australia a meat factory has been established in the ranching country itself. Here the cattle do not have to travel so far on the hoof, and aeroplanes replace the stock-route, for the meat is flown to the ports for export.

Beef cattle in the South American *campos* of Brazil and *llanos* of the Orinoco Basin are reared in a similar way.

In some African savannas, e.g. Kenya and the Rhodesias, the European settler owns land and employs black labourers to farm it. In protectorates like Uganda, however, the white man

[*C.O.I.*

East African Ground-nut Scheme

In Tanganyika it was planned to clear 2,000,000 acres of uncultivated, largely tsetse-infested bush to grow food for millions, but results were disappointing. Note the baobabs, which resisted all efforts to uproot them.

acts as adviser to Negro landholders. Kenya's white planters export coffee, cotton, sisal hemp, tea, and tinned pineapples, while tobacco and maize are grown in the Rhodesias. Nyasaland Protectorate produces tea and tobacco, and the people of Uganda grow cotton most successfully. In the savanna portion of the Anglo-Egyptian Sudan British engineers built the famous Sennar Dam, ponding back the Blue Nile in a reservoir which irrigates cotton over an area as large as Wales.

In Central and South Tanganyika millions of pounds have been spent in clearing savanna bush country in an attempt to grow ground-nuts and sunflowers. This scheme, it was hoped, would help to overcome a world shortage of fats. Unfortunately, the white man was in too great a hurry to achieve a victory, and unexpected difficulties led to disappointing results. Nevertheless, with the exercise of care and foresight, plans like the Tanganyika ground-nut scheme can be made to succeed, and are indeed essential if food supplies are to keep pace with the alarming increase in the world's population. Over 2,400,000,000 human beings now swarm over our planet, and each year there are 30,000,000 more mouths to feed, or over 80,000 every day.

Miners in Savanna Lands.—The white man has not neglected the great mineral wealth of tropical continental regions.

In Australia white miners produce silver, lead, zinc, and copper in the savanna interior of Queensland.

In Africa black miners work under white overseers. In many cases they urgently need the money they earn in order to pay their taxes. Often they work under contract for a few years in mines that are sometimes hundreds of miles from their homes. During this period they live in special compounds, and their activities are strictly controlled by regulations. This long separation from home influences, together with the unsettling effects upon African minds of European civilisation, tends to break up tribal life and leads to a deterioration in the character of the black miner.

Tanganyika produces gold and diamonds. Southern Rhodesia supplies much of the world's chrome and asbestos, while fuel from her huge Wankie coalfield travels northwards by rail to smelt copper in Northern Rhodesia and the Katanga Province of the Belgian Congo. The Katanga, one of the richest mining regions in the world, also yields tin and diamonds, and ranks first as a producer of uranium, the source of radium. In Nigerian savannas the Bauchi Plateau is particularly rich in tin.

With the generation of cheap power from a huge hydro-electric plant under construction at Jinja, Uganda's enormous deposits of copper and phosphates will be made available. If the people can be persuaded to use them, these phosphates will do much to save the soil from starvation. With the aid of this power, cotton cloth and cement will be manufactured locally, instead of being imported from Britain, while the flow of the Nile, so important for Egypt's irrigation requirements, will be regulated.

In South America gold and diamonds lured settlers to the Brazilian State of Minas Geraes, or "general mines". Here manganese, mica, gems, and quartz crystals are also produced, and mountains of iron ore are being levelled to the ground at the rate of 1,500,000 tons a year. To the modern pioneer these Brazilian campos offer a future in agriculture, mining, and industry as exciting as America did to the European emigrants of the nineteenth century.

A solid hill of iron ore, recently found in the Guiana Highlands of Venezuela, may prove to be one of the greatest mineral discoveries of the twentieth century. The richest iron-ore deposits of the United States are waning, and ore from this Venezuelan iron mountain, Cerro Bolivar, will be exported to feed the steelworks of Philadelphia, Baltimore, Birmingham, and Pittsburgh.

EXERCISES

1. On an outline map of the world mark by different shadings (i) tropical continental and (ii) temperate continental lands. Describe similarities and differences between the climate and natural vegetation of these two climatic regions.

2. "Nature arms and equips an animal to find its place and living in the earth, and, at the same time, she arms and equips another animal to destroy it." Give examples from various regions, and particularly from savanna lands, to prove the truth of this statement.

3. "There is not enough food for all; Nature scrambles what there is among the crowd." In what parts of the world does this rule still apply? Show how in different climatic regions man is trying to make food supplies keep pace with the world's rapidly increasing population.

4. "Nomads were the terror of all whom the soil, or the advantages of a market, had induced to build towns." Quoting examples from different regions, explain how this statement is true of former times and why such raids by nomads upon agriculturists and merchants are nowadays less frequent.

5. About 1,900 years ago the Roman author, Pliny, said: "Always something new out of Africa." Show how this statement is still true.

CHAPTER XXVII

The Hot Belt—5. Equatorial Lands

WORLD POSITION AND CLIMATE

Temperature.—Stretching from about 5° N. to 5° S., equatorial regions, as shown in Figs. 73 and 77A, straddle the Equator. Very hot weather prevails throughout the year, for the noon-day sun is always high and twice a year is overhead. The daily range of temperature is greater than the annual range, which is smaller here than anywhere else on earth (Fig. 80). In fact, after the day-time heat night can, by contrast, be described as "the winter of equatorial regions", although, with cloudy skies checking loss of heat after sunset, it is muggy and anything but cold.

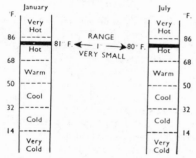

Station : JALIUT (Marshall Islands—Pacific)

TEMPERATURE

Fig. 80.

PRECIPITATION

Annual Amount : Very heavy (176·5″)
Seasonal Distribution : All seasons, with two maxima, one following each of the two periods of the overhead noon sun

Nov. to April 49%   51% May to Oct.
(Winter and Summer cannot be distinguished)

Doldrums low-pressure belt at all seasons—convectional thunderstorms

Precipitation.—Over these lands the Doldrums low-pressure belt of calms is in control at all seasons. Following a fine morning, thunder-clouds gather and by late afternoon tower up to a climax in convectional storms. Amidst torrential rain "the wind and thunder roar their loudest, and the lightnings smite the earth with intolerable light, frightening the poor monkeys in their wet, leafy habitations" (W. H. Hudson).

These thunderstorms bring heavy rain at all seasons (Fig. 80), although short dry spells sometimes occur. They are particularly severe shortly after the two periods when the noon sun stands overhead, for then the air is hotter than usual and convectional currents grow more powerful.

The cloudy skies and heavy rains prevent day-time temperatures from rising as high as they do during summer in cloudless hot deserts. Nevertheless, the white man finds this damp equatorial heat more oppressive than the dry heat of deserts. Shade gives no relief, everything is soon shrouded in mould, and even tinned food quickly goes bad.

Natural Vegetation and Animals

Describing the equatorial Amazon Basin, H. G. Wells writes:

"This forest was interminable, it had an air of being invincible, and man seemed at best an infrequent precarious intruder. One travelled for miles amidst the still, silent struggle of giant trees, of strangulating creepers, of assertive flowers. Everywhere the alligator, the turtle, and endless varieties of birds and insects seemed at home—but man, man at most held a footing upon resentful clearings, fought weeds, fought beasts and insects for the barest foothold, fell a prey to snake and beast, insect and fever, and was presently carried away. The puma, the jaguar, were the masters here."

In this equatorial hot-house atmosphere the "law of the jungle" rules. Thousands of different species of plants struggle in a free-for-all fight to reach sunlight, forming a "universal verdure of bushes, creepers, and trees—trees beyond trees, trees towering above trees". The tallest cover the rest with a "roof-garden" of foliage. Beneath this leafy canopy the prevalent gloom is occasionally relieved by "a golden shaft of light falling through some break in the upper foliage, giving a strange glory to everything it touches—projecting leaves, and beard-like tuft of moss, and snaky bush-rope" (W. H. Hudson).

An aerial view would show a green ocean of vegetation

whose monotony is unbroken, save for vivid splashes of bright orchids and gaps provided by rivers or man-made clearings. Seen from ground-level these rain forests present a less even appearance, for three layers can be discerned:

(*a*) Giant trees tower up to 150 feet or more, supporting their tremendous weight upon woody buttresses that grow out from the base of their trunks.

(*b*) Beneath these monsters grow trees which by comparison seem small, although if found in temperate forests they would rank as tall.

(*c*) Usually far too little light penetrates the leafy roof for a really abundant undergrowth of bushes to flourish. Nevertheless the ground is littered with a tangled mass of roots, creepers, and decayed vegetation that has fallen from above, together with mosses, ferns, and other shade-dwelling plants. Along sunlit riversides and in abandoned clearings, however, dense undergrowth does thrive. To the traveller the most formidable obstacle is the mangrove swamp of the river-bank and sea-coast. From slimy mud the mangrove rises upon stilt-like roots, which at high-water are submerged.

Although these forests are evergreen, one can always find trees whose turn has come to lose worn-out leaves in order to make way for new foliage. With no rhythm of changing seasons, there is no need for seed-time, blossom, and harvest to keep to the calendar. Equatorial fruits therefore ripen haphazardly throughout the year and can be picked at any time.

Mahogany, ebony, rosewood, greenheart, and numerous other hardwoods abound, but so intermingled are the various plants that to find several of a kind close together is the exception rather than the rule. Countless *parasites* cling to their hosts, as their victims are called, sinking their roots into the sap that nourishes them. The word "parasite" comes from the Greek *para* (beside) and *sitos* (food), and means "eating beside, or at the table of, another". *Epiphytes*, e.g. orchids, swarm over the branches of trees. An epiphyte (Greek *epi*=upon, *phyte*=plant) is a plant which grows upon the outside of another one but which, unlike a parasite, does not prey upon it for food. Festoons of rope-like *lianas*, or creepers, grip tree-

[*I.C.I.*]

A Problem of Tropical Africa

Cattle are reared in the Equatorial Highlands of Africa. This cow is suffering from nagana, caused by parasites called trypanosomes. These live in the blood of antelopes and buffaloes, which are unharmed. The bite of the tsetse fly transmits the parasites to cattle and other animals. Game preservation thus encourages nagana. Clearing the bush destroys the tsetse but may lead to soil erosion.

trunks in a stranglehold or drape themselves from branches. Dead or rotting vegetation provides sustenance for *saprophytes* (Greek *sapro* = decayed, *phyte* = plant), the scavengers of the plant world, which feed upon decayed matter.

The luxuriance of the flora of the equatorial rain forest is rivalled by the abundance of its fauna. Monkeys and gaily-coloured parrots and birds of paradise chatter noisily in the tree-tops. In marked contrast to this leafy clamour, silence reigns at ground-level, where vegetation barriers obstruct the passage of almost all but the heavy elephant, the slithering snake, and the lithe big cat.

Insects fill the air and swarm over the ground; most irritate man and many endanger his health. Mosquitoes by the million breed in marshes. Germs injected by a mosquito's bite into man's blood-stream cause malaria and yellow fever. Every year some 2,000,000 people die of malaria, and we in Britain should be grateful for our cool climate, in which these deadly germs cannot live. For ever waging war against tropical diseases, the white man drains the swampy breeding-grounds or covers them with oil, which both suffocates and poisons the larvae of the mosquito. Interference by man in Nature's realm, however, sometimes leads to unexpected and unfortunate results. In some regions, for instance, the practice of oiling swamps has been abandoned, for oil kills not only man's enemies but also his allies, e.g. toads, which prey on beetles that damage sugar-cane. Instead of pouring oil on ponds in such areas, minnows which feed upon the larvae of the mosquito are placed into them. Native huts and the dwellings of the white man are also fumigated by being sprayed from stirrup pumps, first with D.D.T. mixed with paraffin and afterwards with benzene hexachloride.

Termites and tsetse flies also abound. Yet another pest is the hookworm, which bores into the bare feet of the victim, enters his blood-stream, and finally lays its eggs in his intestines. Much of the laziness of the inhabitants of hot lands can be traced to "miner's anaemia", a disease caused by this parasite, and to other tropical diseases such as sleeping sickness and bilharzia.

Man in Equatorial Regions

Some equatorial peoples work for themselves to get a living. They include primitive hunters and collectors; primitive agriculturists; cash-crop farmers; and South Sea Islanders. Others, such as plantation-labourers, lumberjacks, and miners, work for the white man who has provided money and machinery to tap the wealth of forest and mine, but who finds manual work impossible in this forbidding climate.

Primitive Hunters and Collectors.—In the Ituri Forest of the Congo Basin lives the dwarfish pygmy, a hunter and collector untouched by civilisation.

He shelters from heavy rains in a rude hut of leaves and branches, which is abandoned when the family group moves on.

270

He kills his prey with a bow and poisoned arrows, or traps it in a camouflaged pit. Fish are caught in a variety of nets and traps, or are diverted from the river into the poisoned waters of a shallow side-channel. The women gather roots, berries, nuts, fruits, and honey, while not even grubs are beneath their notice.

With neighbouring villagers less primitive than himself the pygmy trades the flesh of his prey for crops, especially bananas, of which he is exceedingly fond. Being timid, he stealthily helps himself to whatever he fancies and leaves meat in payment. Conversely, should the villagers want meat, they know that by leaving bananas at an agreed "market-place" they will, on their return later, find a fair exchange awaiting them. This dumb barter, transacted without either side seeing or meeting the other, is called *silent trade*.

If, in the future, civilisation invades his forest fortress and shatters his customary way of life, the pygmy, like other primitive peoples, will probably die out.

In remote parts of Malayan forests the Semangs and Sakais resemble pygmies in their dependence upon hunting. Unable to count beyond three or four, they are backward and, like the pygmy and the Bushman, are regarded by scientists as "museum-pieces". Similar hunters and collectors roam the *selvas*, as equatorial forests are called in the Amazon Basin.

Primitive Agriculturists.—Taking advantage of a short dry season which brings a welcome change from heavy rains, the Fang people of the Congo Basin cut down and burn all but the tallest trees. In a soil fertilised by the ashes and dead leaves they plant manioc, yams, and bananas. Weeding, a never-ending task, is left to the women, while from riverside camps the men fish, hunt animals, and collect nuts and fruits.

After five or six years of leaching by heavy rains the soil is impoverished, and the whole community moves on to repeat the wasteful process. Subsistence farming by this destructive "slash-and-burn" method of shifting agriculture resembles that practised by savanna crop-growers.

In the Amazon Basin .many tribes combine hunting with

shifting agriculture. The men live the exciting life of the hunter, killing animals with their blow-pipes. Among some tribes the secret of making deadly poison for their darts is so closely guarded that even the hut that sees its manufacture is burned down. The monotonous task of growing manioc, rice, and bananas is given to women, as befitting their low position in primitive society. Manioc, the chief crop, keeps longer than other foodstuffs in this hot, wet climate. Its starchy roots, once the poisonous juice has been squeezed out, yields tapioca and cassava flour.

Amazonian peoples are very backward. Some scorn shelter but others build huts in clearings which, well removed from rivers, are hidden from enemies who might pass by in their canoes. Within the palm-thatched hut a hammock of palm-leaf fibres is slung well above the damp and insect-infested ground. The huts themselves are usually perched upon poles.

The easily satisfied needs of primitive equatorial peoples for clothing are met by tree-bark or by skins. Often their bodies are painted or tattooed and are adorned by feather head-dresses.

Cash-crop Farmers.—By growing crops to satisfy the demands of temperate lands for tropical products, the cash-crop farmer can settle permanently instead of wandering, and can enjoy a rather higher standard of living than either the primitive hunter or the shifting agriculturist. He sells his cacao or his palm-oil to European and North American customers, and consequently can afford to buy manufactured goods which otherwise he would have to do without.

Nearly half of the world's cocoa comes from the Gold Coast. Here are to be found not only the heat and rain needed by the cacao tree but also the calms wherein the pods, which grow straight from the trunk and main branches, run no risk of being snapped off by strong winds. The pod is lopped off with a knife and the two beans embedded within it are removed. After fermenting beneath banana leaves, the beans are dried upon trays.

Palm-oil is to Nigeria what cacao is to the Gold Coast. Looping himself by a rope to the trunk of the giant oil-palm, the Negro hitches himself aloft to pick the fruit, from which palm-oil is crushed. The oil is exported, particularly to feed Merseyside soap factories and South Wales tinplate works. Palm-oil and the equally important palm-kernel oil are also used to make margarine.

Cash-crop cultivation has its dangers. Bumper crops can bring poverty to their growers by causing selling prices to slump to disastrously low levels. Again, disease may ravage the crops. In the Gold Coast the swollen shoot and black pod diseases have ruined millions of cacao trees. At present the only way of fighting swollen shoot is to uproot the infected trees. Nevertheless, scientists are hoping to save the cacao trade of West Africa from doom by finding parasites or insecticides to kill the mealybugs responsible for this disease.

South-Sea Islanders.—The inhabitants of the equatorial South Sea Islands of the Pacific Ocean resemble those of mainlands in the darkness of their skin, the scantiness of their clothing, and the flimsiness of their huts. They live on fish, bananas, breadfruit, yams, and mangoes. Their mainstay, however, is the coconut palm, which to Pacific Islanders is as indispensable as rice is to the Chinese or as the reindeer is to the Lapp. It provides them with food, milk, oil with which to anoint themselves, and fibres and leaves for clothing and shelter. Coconuts are also used as currency and for gifts.

Periodically the white trader in his schooner visits these islands to take aboard the copra that Islanders have stored. Copra is the dried, white kernel of the coconut. Manufactured goods, which fascinate the Islander, are given in payment. The copra is exported to temperate lands to be made into confectionery, margarine, and soap. From coir, the fibre of the outer husk, ropes and matting are woven.

Plantation Workers.—The haphazard distribution of trees of the same species seriously hampers the collection of equatorial forest products. When, for instance, two rubber trees are widely separated much time is wasted in getting through the dense forest from one to the other. Throughout the nineteenth century the Amazon Basin of Brazil easily led the world in producing rubber. It would sometimes take the *seringueiro*, or rubber gatherer, several hours to hack his way from one tree to the next, and he could scarcely be blamed for bleeding the nearest trees until they would yield no more. In 1876 thousands of rubber seeds were smuggled out of Brazil and temporarily housed and tested in Kew Gardens. The survivors were shipped to Ceylon and there planted out in specially prepared clearings. Similar plantations were later established in Malaya and the East Indies, and by 1913 Brazil had lost her supremacy.

Instead of following the example of the reckless Brazilian seringueiro, the Asian coolie at sunrise places a cup beneath a slight gash made in the bark. By noon he has collected hundreds of small quantities, and in a short time the trees recover.

Bananas, sugar-cane, pineapples, sisal hemp, cacao, and coffee are also grown in plantations. Only by these common-sense methods can the increasing demands of industrialised temperate lands for tropical foodstuffs and raw materials be fully satisfied.

The preliminary task of clearing the forest is both difficult and costly, and so far it has been undertaken only by the white man. He possesses the knowledge and skill, the money and machinery by which the forest, age-old master of primitive peoples, has at last been partly tamed. Large plantations have been established by the British, French, Dutch, and Belgians, in equatorial regions that they control—or, in some cases, did control until quite recently.

In 1927 the Ford Company of America obtained rights to clear a huge tract of forest in the Amazon Basin in order to plant rubber trees. A new town, Fordlandia, was built, complete with shops, sawmills, clubs, garages, sports-grounds, and a railway, power station, hospital, cinema, and wireless station. The health of plantation workers was improved. Often against their will, they were persuaded to abandon their monotonous diet of beans, manioc, and rice in favour of modern dishes rich in vitamins, and to wear sandals to keep their feet free from hookworms. Partly because of disease among the trees,

[*C.O.I.*

A Rubber Plantation in Malaya

Each division of this large plantation has its own little village for the workers. Why is the type of house shown suitable for an equatorial climate?

the scheme was abandoned by the Americans in 1946 and this outpost of modern civilisation in an equatorial forest was sold to the Brazilian Government, which is attempting to maintain it. Similar rubber plantations under American control have been successfully established by the Firestone Company in Liberia.

The white man's energy is so sapped by the hot, wet climate that he can do no more than become an overseer of dark-skinned workers, who, being plentiful, provide a source of cheap labour. Most tropical peoples feel little discomfort in this hot-house atmosphere, for they are adapted to withstand its harmful effects (page 76). In tropical lands the white overseer, doctor, engineer, or Government official is paid a high salary and, to safeguard his health, is granted long periods of leave in his temperate homeland. Modern drugs are playing a great part in the conquest of tropical diseases to make the

275

hot belt safer for mankind. Paludrine, for instance, protects man from malaria and cures him if he is already suffering from this centuries-old scourge of humanity.

The introduction of plantation agriculture into equatorial regions bristles with problems. Keeping weeds under control is a never-ending task. Disease both in the plantations and in their human attendants must be kept at bay. Years must elapse before newly planted trees yield profits. Moreover, there is always the risk, so common to *monoculture*, or single-crop cultivation, that supply may outstrip demand, when prices will slump to ruinous levels. Again, only when the worker is treated well can he be expected to work satisfactorily. Food and shelter must be provided for him, his health must be attended to, and he should be given the opportunity of receiving at least a smattering of education. Such care for his welfare costs money. Finally, torrential rains and the quick-growing forest hamper the building and maintenance of the roads and railways by which plantation products are carried to the rivers and ports for export.

Lumbermen and Miners.—Three difficulties face the equatorial lumberman—to find the tree, to fell it, and to float it. In the tangled maze of intermingled species it is none too easy to find a suitable specimen of the particular kind of tree he wants. Having made his choice, he must then build a platform round the trunk before felling can proceed at a level above the gnarled buttresses that support the tree. Moreover, undergrowth and smaller trees must be cut down to allow the forest giant to crash to the ground. After hauling them with difficulty through swampy jungle, the lumberman finds only too often that the logs will not float, and that they must be taken to the port on a cradle, or raft. It is not surprising that lumbering in equatorial forests lags far behind that in the temperate and cold belts.

Only where minerals are easily accessible or are of exceptional value does mining take place. Here, as in most of the hot belt, the white overseer supervises coloured miners. In Malaya and the East Indies rich supplies of tin, cheaply worked by Asian labour, have helped to close some of Cornwall's tin mines. From the East Indies comes oil, while the Gold Coast exports gold, diamonds, bauxite, and manganese.

TROPICAL AND EQUATORIAL HIGHLANDS

THE highlands of the hot belt rise as cooler islands above the sea of trees that engulfs the sweltering lowlands (Fig. 81). In climbing equatorial mountains of about three miles in height one passes through a natural vegetation that changes with the changing climate as much as it does in a sea-level journey of 6,000 miles from Equator to Poles. From dense equatorial forests on the lower slopes, the mountaineer travels in turn through temperate deciduous forest and coniferous forest until, beyond the tree-line, he emerges into temperate grassland, above which tundra leads up to snow-capped peaks.

In South America the thin atmosphere of the Andes hampers work and endangers health. For instance, no train on the Central Railway of Peru is without its doctor, nurse, and oxygen supply.

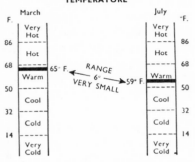

Station : NAIROBI (Kenya)—5,450 feet above sea-level

Note that the hottest month is March, one of the two periods of overhead noon sun and just before very heavy rains of a monsoon nature in April (8·3″)

PRECIPITATION

Annual Amount : Moderate (39·2″)

Seasonal Distribution : All seasons, with two maxima following the periods of the overhead sun.

Nov. to April 70%   30% May to Oct

(Summer and Winter cannot be distinguished

Doldrums low-pressure belt at all seasons—convectional rains, but these are less heavy than in equatorial lowlands

Fig. 81.

Wherever railways are absent in Andean countries the sure-footed but surly llama carries man's goods. Nevertheless, in spite of all the obstacles to progress, the discovery of valuable minerals has led man to open up these and other tropical highlands. In tin-mining Bolivia ranks second to Malaya, while Colombian mines yield platinum, emeralds, and gold. In the East Indies machinery bound for gold mines in the mountainous heart of New Guinea was originally flown in by aeroplane.

277

The comparatively cool climate of highlands in the hot belt permits the white settler to enjoy good health, although the strong sunlight encourages his children to grow rapidly beyond their strength. In the East Indies Dutch merchants recuperate in their mountain retreats from the sultry heat of the ports, while on Kenya's plateaux British colonists farm great estates. In tropical South America descendants of sixteenth-century Spanish conquerors live without discomfort in the Andes. In fact, in the high Bolivian plateaux the enemy to health is not heat but the icy "Harvest of Death", a wind that causes lung disease. Andean Indians keep themselves warm by wearing shawls and woollen masks and by basking in sunshine, shunning like the plague the icily cold shade. Night temperatures in this rarefied atmosphere are often well below freezing-point.

EXERCISES

1. You are placed on a large island in mid-ocean and are given twelve months in which to find out whether you are at the Equator or at the Tropic of Capricorn (23½° S.). Explain how you would be helped by observations of (i) the sun's apparent movements, (ii) the climate, and (iii) the vegetation.

2. "Until Science solves the problem of tropical disease, East and West Africa must not be looked upon as an area for white colonisation. Perhaps they will never be a white man's country in any real sense" (Field-Marshal Smuts). Discuss this statement.

3. Subject for debate: "The coming of the European to Africa has done more harm than good."

4. (a) Within the Tropics the density of population in the Andes is much greater than on the adjoining lowlands, whereas in the temperate belt the reverse is the case. Account for these facts. (b) How have the Andes both helped and hindered economic development in South America?

CHAPTER XXVIII

Conclusion

WE have reached the end of our study of climate and its effects upon life, although much has been left unsaid. Summing up, we see that solar energy, by setting in motion the atmospheric "boiler", provides warmth and rain to sustain over the earth's surface a layer of plant, animal, and human life. We see that man, the highest form of life, ever seeks to master his surroundings by his skill and science but that, despite his many spectacular triumphs, Nature often has the last word.

Man is physically and mentally most alert in temperate lands and is least energetic in the hot and very cold belts. Of course, in overcrowded hot lands like South-east Asia pressure of population forces him to overcome his tendency to become languid, and in such regions he must work hard—or else starve. That mental vigour is brought about by a favourable climate is also emphasised by the high value set upon education in temperate lands, especially in Western Europe and North America (Fig. 82). Moreover, if possession of motor cars and lorries can be taken as a sign of progress and prosperity, Fig. 83 shows that these same favoured regions lead the world in standards of living. Maps showing the world distribution of radio and television sets, telephones, etc., would reveal the same kind of pattern.

Briefly reviewing the climatic regions in turn, we see man in the very cold belt fighting against powerful odds and with difficulty wresting a living from forbidding tundras, although here and there scientific progress gives him visions of a future somewhat less grim.

Similarly, hopes of the taiga of the cold belt ranking among the world's great food storehouses are slender, although in parts farming is possible. Here, too, man labours against climatic

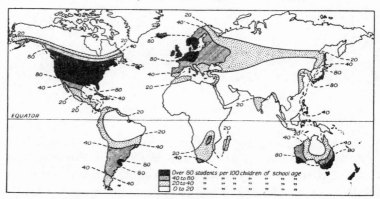

Fig. 82.—*Number of Students (of all ages) in Schools and Colleges for every* 100 *Children of School Age*

odds. Nevertheless, from these huge forest lands come vast quantities of timber, paper, furs, and minerals.

In cool temperate margins we find that climate becomes man's friend rather than his foe. It rouses him to give of his best, making him work to satisfy his needs, yet allowing him sufficient leisure to enjoy the fruits of his labour and to cultivate his mind. On the whole, his standard of living is high. Here are many of the world's greatest workshops, while hand in hand with industry go farming and fishing. Britain in particular may be called "The Fortunate Isles". Her people, never knowing what to expect next from the ever-changing weather, are kept mentally alert. Endowed with fertile soils and rich in coalfields, she is surrounded by seas teeming with fish and is ideally situated for trade with the whole world.

This hustle and bustle dies down as we move equatorwards into warm temperate margins. Here, in bright and cheering sunshine, man pursues his way in a more leisurely and carefree manner. Nevertheless, some parts, especially in Eastern Asia, are so overpopulated that toil from morning till night can alone save him from starvation.

In temperate continental interiors we see that distance from the sea brings dry conditions, and in some parts only by irrigation or by dry farming can the land be made fruitful. Here the danger of soil erosion looms large. Cool temperate

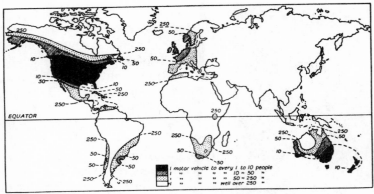

Fig. 83.—Number of Motor Cars compared with Population

continental regions are noted for their remarkably great seasonal change in temperatures. Although the dry atmosphere makes him energetic, the farmer is forced to rest from his labours by the long, cold winter. Equatorwards the cool temperate fade into warm temperate continental lands, where we find that higher temperatures bring about a rather more easy-going way of life.

Finally, in the hot belt, as in the cold and very cold belts, we see man waging a ceaseless war with Nature. A broiling sun does not encourage him to work, but the rainfall is too heavy, or too light, or too unevenly distributed, to make life easy for him. Insect-borne diseases wreck his health and infect his animals. In hot deserts his struggle against drought is eternal. In monsoon lands and in tropical continental savannas summer rains bring salvation, but should they fail he starves. In tropical maritime and in equatorial regions scorching heat and torrential rains join forces against him by giving rise to an ever-encroaching forest, soil exhaustion, disease, and insect pests. Throughout the hot belt poverty is the normal lot of mankind, save where highlands enable the white man to settle or where, as in tropical Australia, the coloured man is debarred.

Man's persistent struggle to master Nature has been vividly described by the nineteenth-century writer and explorer, Winwood Reade. Writing in 1872, he said:

"Thus man has taken into his service, and modified to his

281

[A.P.

Man's Search for Knowledge

This rocket was launched in New Mexico, U.S.A., and reached a height of 135 miles. By such methods man's knowledge of the upper air is increased, and possibly in the future he himself may fly into space.

use, the animals, the plants, the earths, and the stones, the waters and the winds, and the more complex forces of heat, electricity, sunlight, and magnetism, with chemical powers of many kinds. By means of his inventions and discoveries, by means of the arts and trades, and by means of the industry resulting from them, he has raised himself from the condition of serf to the condition of a lord." Then, in a vision of the future, he goes on, "But the Spirit of Knowledge is gradually spreading over the planet and upwards to the skies. . . . And then, the earth being small, mankind will migrate into space, and will cross the airless Saharas which separate planet from planet and sun from sun."

In this book we have seen how, in some parts of the world, man has developed a civilisation in the course of his fight to overcome the drawbacks of his environment, while in other parts the struggle is still being fiercely waged. He may never succeed in fulfilling Winwood Reade's prophecy, but since the above words were written he has achieved marvels. It will be fascinating to watch his progress, for the twentieth century has seen developments undreamt of in past ages. Never before has the future held for mankind such wonderful promise—or such great perils.

282

Index

Aborigine, 57–8, 138, 142, 160, 240
Abyssinia, 102, 221, 228–9, 243
Adaptation to drought, 142–4
Adaptation to environment, 142, 202, 238
Afghanistan, 216
Africa (see also separate countries and regions), 64, 102, 218, 231, 254, 256–64
Agriculturists (or crop-growers), 59, 214, 239, 243–8, 256, 258, 259–62, 270, 271–2
Alberta, 75, 204–5, 209
Aleutian low-pressure belt, 87
Amazon Basin, 267, 271, 274
Anglo-Egyptian Sudan, 228, 263
Animal life (see also Fauna), 137, 156, 166, 176, 186, 202–3, 224, 238–9, 254–5, 269–70
Antarctic Circle, 25, 50
Anticyclone, 117, 124–7, 139
Anti-meridian, 25
Antrycide, 257
Arabia, 52, 216, 236, 245
Arabs (see also Bedouin, Tuaregs), 240–2
Arctic Circle, 20, 25, 47–50, 130, 133, 154
Arctic regions (see also Tundra), 58–9, 62, 125, 152, 155, 159
Argentina, 69, 75, 150, 186, 197, 218
Asia (see also separate countries and regions), 77, 97, 100–1, 112, 133–4, 136, 150, 169, 175–6, 179, 186, 195, 212, 216, 221, 224, 227–8, 231, 236, 253, 279–80
Assam, 222, 225
Atacama and Peruvian Desert, 236
Atmospheric circulation, 12, 94–6, 279
Australia, 75, 97–8, 100, 102, 186, 188–9, 191, 197, 218, 221, 225, 227–8, 234, 236, 239, 249, 262, 281
Axis, 18, 21, 24, 27, 28, 37, 41

Bar, 84
Barysphere, 14
Bay of winter cold, 133–4
Bechuanaland, 218, 258
Bedouin, 58, 240
Belgian Congo, 264

Belgium, 179, 262
Bengal, 222, 225
Benguella Current, 131, 135
Benzene hexachloride, 270
Bilharzia, 255–6, 270
Black pod, 273
Bolivia, 277–8
Boll weevil, 196
Bora, 75
Boulder Dam, 245
Brave West Winds, 93
Brazil, 21, 132, 233, 262, 265, 274
Brazilian Current, 131
Brickfielder, 75
British climate (see Temperate maritime)
British Columbia, 22, 75, 134, 171, 178
British Guiana, 232, 234
British Honduras, 232, 234
British Isles (or Britain), 11–12, 22, 31–2, 35, 37, 40, 44, 47, 51, 81–3, 109, 112, 120, 125–7, 133, 141, 148, 179, 188, 191, 193, 197, 210, 225, 228, 236, 242, 257, 262, 264, 270, 280
Buran, 75
Burma, 224–5, 227
Bushman, 57, 160, 240, 271

California, 75, 191–2, 197
Californian Current, 131, 135
Campos, 262, 265
Canada, 22, 25, 31–2, 53, 68, 154, 159, 165–6, 169–72, 176, 180, 203, 207, 209, 210, 234, 244
Canaries Current, 131, 135
Cancer, Tropic of, 25, 44–5, 133
Canterbury Plains, 75, 176
Capricorn, Tropic of, 21, 25, 44–5
Caribbean coastlands, 231–2
Carnivorous animals, 137, 254
Cash-crop farming, 260–3, 270, 272–3
Central America (see also separate countries), 231
Ceylon, 225, 274
Chile, 107, 134, 179, 191
China, 11, 101, 192–5
China climate (see Eastern warm temperate)

283

Chinese, 56, 77, 186, 192, 194–5, 228, 273
Chinook, 75–6, 107, 207
Climate, 51
— and civilisation, 58–9, 152–3, 187
— and history, 59
— and man, 56–8, 187
— and vegetation, 137–53
— and weather, 51–2
— changes in, 59
— continental, 64–6
— description of, 54–6
— maritime, 66–8
— types of (*see also* separate names), 150, 152
Cloudiness and temperature, 71–4, 105, 220, 222, 237, 266–7
Clouds, Formation of, 103–7, 110–2, 122–3, 125
Cold belt, 148, 150, 152, 165–73, 174, 276, 279, 281
Cold deserts (*see also* Tundra), 150, 156
Cold-water coasts, 106, 220
Cold wave, 124, 126
Cold-weather drought, 143–4, 147
Collective farms, 210–2, 214
Collectives, 163, 169
Colombia, 277
Colorado Desert, 236
Condensation of water-vapour, 65, 72, 75–6, 103–6, 110–11, 122, 136, 230, 238
Conduction, 60–1
Congo Basin, 270–1
Continental air, Temperature of, 63–4, 88, 96
Continental climate, 64–6, 68, 73, 83, 112, 126, 134
Continental "lungs," 96–7, 102, 112, 133, 175
Continental shelf, 180
Convection, 60–1, 107, 109, 110, 125, 267
Convectional rains and thunderstorms, 65, 107, 109, 112, 125, 200, 206, 208, 210, 221, 238, 247, 251–2, 261, 266–7
Cool temperate belt, 150, 174–82, 183, 185, 199–215, 280
Cool temperate continental climate and regions, 150, 174, 184, 199–215, 280
Cooling of air, 105–7
Cornwall, 47, 81, 276
Cuba, 234

Day and night, 28–30, 47–50
— length of, 30, 47–50, 62, 71, 138, 163, 205

Death Valley, 236
Deep South, 195–6, 208
Depressions, 107–8, 110, 112, 114, 117–26, 139, 155, 174–5, 184, 221
Deserts (*see* Cold, Hot, Temperate deserts)
Desert towns, 58, 248
— vegetation, 142–4, 149–50
Dew, 53, 56, 72–3, 103, 105, 125, 238
Doldrums, 87–9, 91, 94–5, 98, 100, 109, 110, 112, 124, 221, 230, 252, 266
Downs (Australia), 202, 218
Dry farming, 205, 218, 280
Dust-bowl, 140
Dutch Guiana, 234

East Africa, 61, 254, 258
East Australian Current, 131
East Indies, 274, 276–8
Eastern cool temperate climate and regions, 68–9, 150, 174–9, 184
Eastern warm temperate climate and regions, 150, 183–6, 192–8, 230
Egypt, 19, 75, 102, 216, 228, 243–5, 248, 256, 264
England, 18, 35, 42, 81–2, 108, 210
Epiphytes, 268
Equator, Definition of, 24
Equatorial climate and regions, 52–3, 58, 61, 64, 72–4, 86, 102, 110, 112, 125, 150, 220–1, 223, 230, 250–1, 253, 266–78, 281
Equatorial Counter Current, 131
Equatorial Highlands, 61, 254, 277–8
Equinox, 45–7, 49–50
Erg, 245
Eskimos, 57, 59, 138, 156–61, 163
Eurasia, 63–4, 96–7, 133–4, 150, 154, 165, 171–2, 191, 200, 214
Europe (*see also* separate countries and regions), 16, 67, 75, 112, 127, 133, 136, 138, 160, 176, 179–80, 182, 187–8, 193, 203–4, 214, 218, 279

Falkland Current, 131
Fang, 271
Fauna, 137, 202, 269
Fellahin, 243–4
Ferrel's Law, 28, 33–6, 91, 93, 130
Finland, 20, 160, 171
Fishing, 137, 157–8, 160, 163, 179–82, 192, 271, 280
Flora, 137, 269
Fog, 53, 103, 105–6, 124, 126–7, 135–6, 180
Föhn, 76, 107
Food chain, 180
Fordlandia, 274

Forests, 145–8
— climatic needs, 145–8
— coniferous, 58, 144, 146–8, 156, 160, 163, 165–6, 170–3, 176, 277
— deciduous, 144, 145–7, 171, 175–6, 277
— equatorial, 58, 137–8, 152, 176, 231, 248, 253–4, 267–9, 271, 273–7, 281
— evergreen, 144, 147, 186, 223, 231, 268
— gallery, 146, 149, 202, 214, 253, 258
— tropical, 186, 231, 281
France, 75, 179, 189, 194
Frontal surface, 119–20
Fronts of depressions, 117
— cold, 120, 122–3
— occluded, 120
— polar, 119–20
— warm, 119–23
Frost, 54, 71–2, 83, 143–4, 165, 176, 186–7, 196, 205, 237
Frost-drainage, 85
Fulani, 259, 261
Furious Fifties, 93
Fur-trappers, 158–9, 161, 166, 168–9

Gambia, 261
Germany, 179, 204
Gila Desert, 236
Gobi Desert, 216
Gold Coast, 272–3, 277
Grasslands, 148–9
— climatic needs, 148–9
— types of (see Campos, Downs, Llanos, Pampas, Prairies, Puztas, Savannas, Steppes, Veld)
Great circle, 22–5, 136
Great Lakes, 179, 200, 208
Great Northern Sea Route, 163, 172
Greenwich, 20, 25, 31–3, 47
Guinea Coast, 75, 102, 221, 252
Guinea Current, 131
Gulf of winter warmth, 133–4
Gulf Stream, 131–2, 134

Hail, 53, 103
Hammada, 246
Hardwoods, 144, 186, 231, 268
Harmattan, 75, 252
Harvest of Death, 278
Hausas, 252, 259, 261
Heat wave, 61, 71, 77, 124–5
Herbivorous animals, 137, 180, 254
Herdsmen (see also Nomads), 59, 163, 212–4, 239–42, 256–61
High Plains, Cattlemen of, 207–8
Hill-stations, 222
Hoar frost, 72
Hoe-cultivators, 260

Hookworm, 270, 274
Horse Latitudes, 87–9, 91–2, 94–6, 98, 100, 102, 107, 110, 114, 117–8, 123, 125–6, 184
Hot belt, 145, 147, 149–50, 220–78, 279, 281
Hot desert climate and regions, 53, 56, 70, 72–4, 103, 105–7, 110, 112, 130, 135, 138, 142–3, 150, 156, 184–6, 220–1, 223, 236–49, 250–3, 259, 267, 281
Hudson's Bay Company trading-posts, 159, 166, 168–9
Humboldt (Peruvian) Current, 131, 135
Hungary, 214
Hunters and collectors, 137, 203
— in deserts, 239–40
— in equatorial forests, 270–2
— in tundra, 156–60
Hydro-electric power, 164, 171–2, 178–9, 234, 264
Hydrosphere, 15

Icebergs, 14, 69, 136
Icelandic low-pressure belt, 87
India, 52, 69, 74, 97–8, 100, 102, 130, 222, 224–7
Insect pests (see also separate names), 152, 192, 203, 206, 218, 224, 257, 269, 281
Interior cool temperate climate (see Cool temperate continental)
Interior warm temperate climate (see Warm temperate continental)
International Date Line, 32–3
Iran, 216, 249
Iraq, 216, 236, 245, 249
Ireland, 81, 83, 176
Irrigation, 58, 188, 205, 215, 218, 224, 227, 243–8, 264, 280
Isobar, 84, 91
Isohyet, 103
Isotherm, 78–83, 84, 133
Italy, 188, 194, 204
Ituri Forest, 270

Japan, 101, 179, 192, 194
Japanese, 77, 194–5, 228

Kalahari Desert, 57, 160, 236, 240
Katanga, 264
Kazakhstan, 214
Kazaks, 212–3
Kenya, 20, 254, 262–3, 278
Khamsin, 75
Khirgiz, 57–8, 212
Knot, 26
Kuro Siwo, 131, 133

Labrador Current, 106, 131, 136
Lamas, 215–6
Lancashire, 82, 107
Land and sea breezes, 90, 97, 106, 237
Lapland, 160
Lapps, 57, 160–3, 273
Latitude, 26–7
Laurentian climate (see Eastern cool temperate)
Leaching, 260, 271
Lianas, 268
Liberia, 275
Light year, 15
Line squall, 123
Lithosphere, 14–5
Little Egypts, 245
Llanos, 262
Locusts, 203, 255
London, 22, 26, 31, 135, 228
Longitude, 26–7, 30–2
Lumbering and lumberjacks, 159, 170–3, 181, 270, 276

Magyars, 214
Malaria, 152, 188, 269–70, 276
Malaya, 271, 274, 276–7
Mangrove swamps, 268
Manitoba, 204
Maritime (oceanic) air, 63
— climates, 66–8
Masai, 258–9
Mealybugs, 273
Mediterranean climate and regions, 114, 126, 150, 183–92
Mercator map, 22–3
Meridian, 20, 25–7, 30–3, 81
Meteorological stations, 53, 127, 163
Mexico, 195, 236
Middle West (U.S.A.), 200, 208
Millibar, 84
Minas Geraes, 265
Miner's anaemia, 270
Mining and miners, 137, 159, 163, 171, 178–9, 197, 209, 215, 218, 234–5, 239, 248–9, 256, 264–5, 270, 276–7, 280
Mist, 85, 105–6, 111, 125–6, 176
Mistral, 75, 189
Mohave Desert, 236
Mongolians, 58, 77
Monoculture, 276
Monsoon climate and regions, 74, 97–102, 112, 145, 150, 216, 220–9, 243, 253, 281
Monsoon winds, 97–102, 130, 221–3, 228
Mosquito, 156, 269–70
Mozambique Current, 131
Murray-Darling Basin, 218

Nagana, 257
Natal, 195, 197
Nautical mile, 26
Negro, 56, 58, 76–7, 196, 231, 233, 260–1, 263
New England, 178
New Guinea, 277
New South Wales, 75, 197–8
New Zealand, 63, 69, 134, 174, 176, 191
Newfoundland, 106, 136, 178–9
Nigeria, 264, 272
Nile, 102, 216, 228, 243–4, 264
— Blue, 228, 263
— White, 228
Nomads (see also Herdsmen), 59, 203, 212–5, 239–42, 248, 256–9, 262
North Africa, 191, 236–7, 242, 251
North America (see also separate countries and regions), 25, 63–4, 96, 112, 134, 136, 150, 154, 156, 165, 170–1, 175–6, 178–9, 200, 209, 210, 279
North Atlantic Drift, 67, 81, 106, 131–2, 134, 136
North Equatorial Current, 131–2
North Pacific Drift, 131, 133, 136
Norther, 75
Norway, 133, 160, 171, 204
Nor'wester, 75
Nova Scotia, 178
Nyasaland, 263

Oasis, 156, 228, 238–40, 242, 245–8
Ocean currents (see also separate names), 67–70, 105–6, 129–32
— and climate, 67–70, 105–6, 131–6, 220, 237
Oceanic (maritime) air, 63
Orbit, 37, 39, 43–4, 48
Oregon, 134, 171
Orinoco Basin, 262
Orographic rain (see Relief rain)
Overgrazing, 139, 204, 257, 260
Oya Siwo, 131, 136

Pack-animals, 195, 228, 239, 255
Painted Desert, 236
Pakistan, 222, 225, 227
Paludrine, 276
Pampas, 185–6, 197, 202, 218
Pampero, 75, 186
Parallels of latitude (definition of), 24–5
Parasites, 255, 257, 268, 270, 273
Patagonia, 69, 107, 150, 176
Path of depression, 120, 122
Pennsylvania, 179
Perspiration and cooling, 76–7, 252
Peru, 135, 245, 277

Peruvian and Atacama Desert, 236
Plankton, 180
Plantations, 231, 270, 273–6
Poland, 179
Polar air (in depressions), 117–24
Polar high-pressure belts, 87, 89, 92, 95, 126
Polar regions, 59, 86, 110
Polar winds, 93–6, 107, 117
Poles, North and South (definition), 21
Poor whites, 196
Population of world, 141, 264
Prairies, 58, 149, 159, 170, 202–8
Precipitation (*or* rainfall), 53, 103–9, 112–4, 122–3, 142, 145–6, 148–9, 155, 165, 175, 200, 221, 237–8, 251–2, 266–7, 281
— annual amount of, 55–6, 116, 155, 165, 200, 237, 251, 254, 267, 281
— form of, 56, 107, 155
— means of, 54
— measurement of, 103
— seasonal distribution of, 56, 112, 116, 155, 165, 174–5, 184, 200, 221–2, 231, 251, 267, 281
— word-scale for, 54–5, 155
Pressure (atmospheric), 84–9
— and altitude, 85
— and rotation, 85–6
— and temperature, 85, 88–9
— belts, 84–9, 112, 114
— gradient, 93, 100
— in depressions, 122–3
— means of, 54
— measurement of, 84
— seasonal changes of, 87–9, 92, 96–100, 102
Puztas, 214
Pygmies, 58, 138, 270–1

Quebec, 171
Queensland, 234–5, 264

Radiation, 60–1, 71–2, 77, 237
Rainfall (*see* Precipitation)
Rain-shadow, 107, 176, 197
Red Indians, 166, 168, 203–4
Reduction to sea-level, 79, 84, 103
Relief rain, 107–8, 112, 175, 184, 221–2, 228, 231, 237, 252
Revolution round sun, 37–50
— and length of daylight, 37, 47–50
— and seasons, 37–43
— and swing of sun, 37, 43–6
— and time, 37–8
Rhodesias, 262–4
Ridge of high pressure, 117, 123–4
Rinderpest, 258
Roaring Forties 93

Rotation of earth, 28–36
— and day and night, 28–30, 47–50
— and Ferrel's Law, 28, 33–6
— and ocean currents, 129–30
— and pressure, 85–7
— and time, 30–3
— speed of, 34–5, 37
Russia (*see also* U.S.S.R.), 65, 126, 130, 160, 169, 204, 210, 244
Ryot, 56, 224, 227

Sahara Desert, 75, 102, 216, 236, 238–40, 242, 245–6, 249, 252, 259
Sakais, 271
Santa Ana, 75
Saprophytes, 269
Saskatchewan, 204–5
Savannas (*see also* Campos, Llanos), 57–8, 137, 142, 145, 149, 242, 253–4, 258–9, 262–4, 271, 281
Scotland, 35, 81–2
Seasons, 38–43
Sectors of depressions, 117, 119–20, 123
Selvas, 271
Semangs, 271
Sennar Dam, 263
Shifting agriculture, 260–1, 271–2
Shrieking Sixties, 93
Siberia, 33, 64, 75, 165, 173
Sirocco, 75
Skin-colour and sunshine, 77
Sleeping sickness, 255, 270
Small circle, 24
Smog, 127
Snow, 53, 74–6, 97, 103, 111, 126, 137–8, 144, 156, 158, 165, 168, 170–1, 174–5, 178, 200, 206, 210
Snow-line, 74
Soil erosion, 139–41, 204, 214, 218, 226, 242, 280
Solstices, 45–7
Sonora Desert, 236
South Africa, 102, 188–9, 197, 218
South America (*see also* separate countries and regions), 63, 112, 130, 174, 185, 202, 231, 262, 265, 277–8
South Equatorial Current, 131, 132
South Sea Islanders, 57, 270, 273
Southerly Burster, 75
Steppes, 58, 149, 172, 202, 215
Subsistence farming, 260, 271
Sunrise and sunset, 29, 46–50, 163
Sweden, 160, 171
Swing of sun, 37, 43–6, 83, 87–8, 96, 112, 114, 116, 125–6, 155, 184
Switzerland, 58, 74, 76, 179
Swollen shoot, 273

INDEX

Taiga, 166, 171–3, 279
Tanganyika, 254, 263–4
Tasmania, 63, 134, 174, 179
Temperate belt, 21, 59, 61, 112, 117, 138, 145, 147, 149, 261, 272–4, 276
Temperate continental climates, 66, 199–200, 251, 280
Temperate deserts, 150, 236
Temperate maritime climate and regions, 67–70, 73, 112, 134, 150, 171, 174–5, 178, 184–6
Temperature, 60–83
— and altitude, 60–1, 79, 220, 254, 277
— and aspect, 74
— and cloudiness, 71–4, 105, 220, 222, 237, 266–7
— and distance from sea, 62–70, 79
— and human body, 76–7
— and latitude, 61–2, 79
— and local winds, 74–6
— and ocean currents, 70, 79, 131–6
— and pressure, 85, 88–9
— and prevailing winds, 66–70, 79
— and water-vapour, 103–5
— annual range of, 54–6, 64, 66, 68, 70, 82–3, 88, 96, 155, 175, 200, 220, 230, 237, 251, 266, 281
— inversion of, 2, 126
— lapse-rate of, 61, 76
— means of, 53–4, 78–9, 81
— word-scales for, 54–5, 154
Termites, 255, 270
Tibetans, 57, 215–6
Ticks, 255
Time, 30–3
— measurement of, 28, 30, 37–8
— standard, 31–2
— sun (local), 30–2
— zones, 32
Trade winds, 67–8, 70, 74, 91–5, 98, 100, 102, 112, 114, 124, 129–30, 135, 150, 184, 221–2, 230–1, 236–7, 252
Transhumance, 160, 191
Transitional regions, 112, 114, 184–6, 250, 253
Transpiration, 142, 144, 146–7, 223
Tree-line, 147, 277
Tropics (see also Cancer, Capricorn), 43–6, 61–2, 66–7, 88, 94, 124, 133
Tropical air in depressions, 117–24
Tropical continental climate and regions, 66, 150, 221, 250–65, 281
Tropical diseases (see also separate names), 152, 269–70, 275–6, 281
Tropical lands, 112, 149, 220–65
Tropical maritime climate and regions, 68, 70, 112, 150, 184, 220–1, 230–5, 236, 250–3, 281
Tsetse fly, 255, 257–8, 270

Tuaregs, 240, 242, 259
Tundra vegetation and regions, 58, 138, 144, 150, 156, 159–60, 163, 165–6, 202, 277, 279

Uganda, 254, 262–4
Union of Soviet Socialist Republics (or Soviet Union), 163, 172–3, 179, 210–5
United States of America, 25, 75, 135, 140–1, 171, 176, 185, 195–7, 200, 203, 207, 209, 210, 215, 234, 236, 239, 244–5, 249, 265, 274
Uruguay, 186, 197

Variables, 67, 70, 74, 92–5, 97, 107–8, 112, 114, 117, 124, 130, 133, 174, 184
Vegetation (see also Desert, Forest, Grassland), 58, 110, 112, 137–52, 156, 165–6, 171, 175–6, 185–6, 200, 202, 223–4, 231, 236, 238, 253–4, 267–9, 277
Veld, 202, 218
Venezuela, 235, 265
Very cold belt, 150, 154–64, 279

Wales, 210, 263, 273
Warm temperate belt, 150, 183, 185, 199, 220, 280
Warm temperate continental climate and regions, 150, 174, 183–4, 199–219, 281
Water-vapour, 65, 72, 75–6, 103–5, 107, 111, 132, 155, 230–1, 238
Weather, 51–2
— and climate, 51–2
— in depressions, 122–3
West Australian Current, 131, 135
West Wind Drift, 131, 134
Western cool temperate climate (see Temperate maritime)
Western warm temperate climate (see Mediterranean)
White settlers in Tropics, 61, 256, 262–3, 278, 281
Willy-willies, 228
Winds (see also separate names), 74–6, 84, 90–102
— and Ferrel's Law, 91
— local, 74–6, 186
— prevailing (planetary), 67, 68, 74, 93, 112, 114, 129, 131–3, 236
— seasonal changes in, 96–102, 112, 114
Word-scales for climate, 54–6

Year, measurement of, 37–8
Yellow fever, 269

288

OCT